DRAIN THE DEFENCE SWAMP

A BLUEPRINT FOR WEAPONS ACQUISITION
REFORM

GARY D. STEWART

STEWART PUBLISHING

Title: DRAIN THE DEFENCE SWAMP

Published in 2020 by Stewart Publishing.

A division of Kaikaku Corporation Pty Ltd.

First Edition 2020.

Copyright © 2020 by Gary D. Stewart.

All Rights Reserved.

The moral rights of the author have been asserted

All inquiries should be made via the website www.DraintheSwampbooks.com.

Printed in Australia by Minuteman Press.

Cover Design by Infinite Designs. Image from Shutterstock

Editing and Formatting by Hilton Copywriting

Disclaimer:

ISBN:978 0 6489524 0 4 (Print)

978 0 6489524 1 1 (eBook)

Visit: www.DraintheSwampbooks.com

DRAIN THE DEFENCE SWAMP
A BLUEPRINT FOR WEAPONS ACQUISITION REFORM

How to FIX every Product Development to be more Affordable, Producible and Problem-Free

ABOUT THE 'DRAIN THE SWAMP' SERIES...

This is the first in a series of four books.

While each book stands alone, collectively they all share a common theme.

That common theme is about a way of thinking that is 'wholly unfit for purpose'.

Each book in turn will propose a reform methodology that is 'wholly fit for purpose'.

Book#1 - Drain the Defence Swamp: A Blueprint for Weapons Systems Acquisition Reform

This first book describes a Defence Weapons Acquisition methodology that is 'wholly unfit for purpose'.

It is wholly unfit for purpose because, despite numerous reviews and reorganisations, Defence Weapons Acquisition programs continue year after year to deliver completely unaffordable results from cost blowouts and long timeline delays that ultimately fail to deliver the promised capability into the hands of our Warfighters.

Affordability and producibility are far too rarely a requirement in Weapons Acquisition programs.

'This is as good as it gets' is the common mantra one consistently hears from Defence.

Obfuscation of both targets and results is the norm, therefore failure to meet set targets, and deliver good program results is the norm.

Defence Weapons Systems Acquisition is a virulent swamp, wholly unfit for purpose.

Book #2 - Drain the Innovation Swamp: A Blueprint for Innovation System Reform

The second book will depict the current Innovation system as 'wholly unfit for purpose' and will outline a reformed Innovation system that is 'wholly fit for purpose'.

Book #3 - Drain the Bureaucracy Swamp: A Blueprint for Bureaucracy System Reform (actual title to be finalised)

This third book will depict a system of political Bureaucracy that is 'wholly unfit for purpose' and will outline a reformed system of Government Bureaucracy that is 'wholly fit for purpose'.

Book #4 - Bold Australia: A vision for a Bold and Daring Future (actual title to be finalised)

The fourth will depict how a 'return to normal' is 'wholly unfit for purpose' for Australia's future and will outline a bold and daring way forward for Australia that is 'wholly fit for purpose'.

WARNING

This book bluntly and brutally talks about an 'Organisational Incompetence'.

As such it will make extremely uncomfortable reading for some.

But Organisational Incompetence is never the fault of any single person trapped within its grasp. Rather Organisational Incompetence is a team sport that requires a passive acceptance of mediocre outcomes by many over long periods of time.

Such deeply entrenched Organisational Incompetence can never be fixed using polite excuses and simple short-term fixes.

Only a brutal and direct confrontation of every fundamental issue or excuse that drives Organisational Incompetence will FIX the underlying problems permanently.

This book takes such a fierce direct confrontational approach.

If you are of a genteel nature uncomfortable with brutal confrontation, then perhaps you should not read this book.

CONTENTS

FOREWORD

There is no Foreword.

At least not like any foreword you would expect. That's because no erstwhile expert or luminary in this field dares comment.

Defence by its very nature is about discipline and conformance - as such it does not tolerate dissent or criticism well.

During my years working in and around Defence I always understood there was a basic fear of retribution preventing anyone from speaking out.

But, as this book came together I realised that I had underestimated that fear.

FEAR is the wrong word.

TERROR is the right word.

People are far too terrified of being cut-off from the flowing rivers of Taxpayer Gold or ostracised from the camaraderie of their mates to put their head above the parapet.

Whether that terror is real or imagined is irrelevant.

However it is formed - it is an extremely effective barrier that stifles all discussion and kills all dissent and reform dead in its tracks.

Silence is the result.

So, there is no Foreword.

There are no Endorsements.

There are no Defence book reviews.

There is just this book.

In Silence.

Alone.

PREFACE

It is only possible for this book to be published in 2020 now that I have
fully retired from all work with Defence.
It would not be possible to expect that having laid bare the absolute
dysfunction and incompetence that is Defence Weapons Acquisition that
I could expect to continue to work with Defence.
Defence is far too vindictive for that.
Having now retired from all Defence work,
I am now free of any, and all such retribution.
I am free to expose the nakedness of both
the Emperor, and his Sacred Cows.
This book is the result.

The closer I became to Defence, the more I discovered that there is no
fact that I could present that cannot be countered with another fact of
Defence's choosing. This is a huge high stakes game with vast Taxpayer
treasure at stake for many very powerful players. In this game, the facts
are not the facts. They are alternate realities.

This book is **not** written for Defence – it is a book written for everyone
but Defence.

The intent of this book is to begin the reform of Weapons Acquisition via
disruption of the Status Quo.

I personally don't have any expectation that anything in Defence will change as a result of this book.

If Defence is true to form, they will simply bunker down and wait out the storm, just as they have always done.

If Defence is true to form, they will paint this book as being just another crack-pot, weirdo, disgruntled contractor, or wacko with an axe to grind, who is peddling half-truths, misinformation, or rumours using someone else's perceptions of reality that have long-since been debunked by Defence. This is the Defence Way.

Whether there is truth in that is not for me to decide. That is for you the reader alone to decide. My job in this book is to lay out the issues and facts as I see them. Then paint some alternate realities as Defence sees them. And then let you, the reader, decide for yourself among these alternate realities where the best truth lies.

Acknowledgements

I have to thank all of those Warfighters, Defence Acquisition personnel, or Defence supplier personnel who helped this interloper from outside of Defence to learn and understand the mysteries and intricacies of the Defence Way. For their own reasons they wanted to remain anonymous. Never to be quoted. But always open and willing to talk and answer my myriad of questions one-on-one using Chatham House Rules.

Chatham House Rules: *any meeting, where participants are free to use the information received, but neither the identity nor the affiliation of any speaker, nor that of any other participant, may be revealed.*

INTRODUCTION

This is a Truly **Brutal** Book.

It is a Brutally Frank Book.

It Brutally reveals the often perplexing reasons behind the dysfunction, incompetence and ineptitude displayed in so many Defence Weapons Systems Acquisition Programs.

It Brutally debunks and destroys the 'this is as good as it gets' mentality that underpins so much of Defence Weapons Systems Acquisition Programs today.

It Brutally depicts the current Weapons Systems Acquisition methodology as **wholly *unfit* for purpose**.

It Brutally portrays a virulent 'Defence Swamp'.

You, the reader, will be appalled at such wilful and blatant waste of taxpayers' hard-earned money.

If you are not appalled, then it will only be because you have become so 'conditioned' to Defence's myriad of excuses that 'this is as good as it gets', that you no longer care about the outrageous waste of taxpayer money.

This book will also propose a robust solution and a new acquisition methodology that is wholly fit for purpose.

Linkages:

While this book is very specifically focussed on Defence Weapons Acquisition problems, in reality the cost and time blowouts, wicked problems, highly dysfunctional outcomes and just plain dumb decisions described in this book also apply across most of Industry, the Corporate world, and other Government functions or Qangos.

In short, what follows is not wholly unique to Defence - but can apply everywhere.

Spaced throughout this book are challenges to you the reader to consider how all this relates to your own personal experiences at your own work or perhaps how you are treated as a customer.

I am sure many readers will experience a sense of déjà vu as this book unfolds.

I am sure some could even re-write this book substituting my Defence examples for their own industry and that book would make complete sense - such is the universal nature of the problems described in this book.

I am sure that astute readers will realise that the content in this book has far wider reform application across many expanses of the economy. The content from both halves of this book can apply just as equally to all types of industries & firms involved in Innovation, Product Design, Manufacturing, or the delivery of Services.

I will return to this theme in Appendix A: where I will outline a few industries that need similar reform.

The First WHY?

Defence Weapons Systems Acquisition Programs have been on the high risk list for many years – simply because, despite numerous efforts at Weapons Systems Program Acquisition Reform since the Collins Submarine development debacle, far too many acquisition programs still experience cost overruns, schedule delays, or operational capability and capability performance shortfalls.

These cost overruns, missed deadlines and failed programs are only symptoms of the problem, not root causes.

Until the actual root causes of these persistent problems are addressed, then no amount of additional oversight, extra regulation, rearranging of organisation boxes, creation of new offices or structures, or changes to processes will help to improve the outcomes of current or future Weapons Acquisition programs.

In truth they will most likely make things even worse.

WHY?

All of these past problems in Weapons Acquisition programs share a common dynamic: moving forward with programs before all of the correct and necessary data and knowledge, needed to make the right decision, is available at every single program decision point, and at the very time it is required.

Today's constrained National budgets will demand a culture of far better decision making across all future Defence Weapons Acquisition programs in order to deliver better and more affordable weapons to our Warfighters.

I have written this book to propose a viable solution to the dilemma in current Defence Weapons procurement, in both acquisition and sustainment, because most Defence Weapons Acquisition programs experience:

- Program budgets that cost way too much initially – and/or even further blowout costs during development
- Program timelines that are far too long – and/or even further blowout timelines during development
- Programs that have far too many 'problems' - either in development – or passed on to the Warfighter
- Programs of very dubious risk mitigation - that deliver poor equipment capability outcomes
- Programs that become 'projects of concern' - and must be remediated with further cost & time overruns

These issues typically result in delivered outcomes where much needed Warfighter capability is compromised – as programs run on and on - or exorbitant costs cause fewer platforms or weapons to be ordered – or far more money needs to be spent on sustainment than forecast – or programs need to be remediated before they are even finished in the Engineering & Manufacturing Development (EMD) phase (before first production).

Taken together, all these current Weapons Acquisition failures constitute a virulent Defence Swamp, wholly unfit for purpose.

This Book is in 2 parts.

Part 1 - The Defence Swamp

Chapters 1 through 13 outline the vital need for reform.

Part 2 - How to Drain the Defence Swamp.

Chapters 14 through 22 outline a bold and daring future Weapons Systems Acquisition Reform methodology.

This book concentrates on Weapons Systems only.

This book makes no comment about the Military Operations side of Defence. It does, however, comment on the impact to our Warfighters from poor weapons development programs.

In general industry, the focus is on the customer as the driver of business strategy. However, in Defence the focus must ALWAYS be on the needs of the Warfighter to successfully prosecute any war.

This book concentrates on Weapons Acquisition only.

This book makes no comment about the operational capability of the various Weapon Systems. That is, it makes no comment about whether the Weapons System is the right choice of capability in the first place.

It focuses solely on the failure modes in acquisition methods used by Defence and the Defence supply chain to design, develop, manufacture and deliver weapons systems to the Warfighter.

This book - and the setting of standards (or benchmarks, or baselines):

In order to understand delivered Weapons Acquisition Program results and performance we first need to understand what type of standard or benchmark those acquisition programs must be able to meet.

Some common options include:

No Standard: No assessment of success or failure can be made, as there is nothing to measure against.

Weak or Low Standard: Then almost every activity can get a pass - no matter how good or bad it is.

Tough or Theoretical Standard: Almost no-one will be able to achieve in the real world (*Breakthrough needed*)

World Best Practice Standard: Some others have already delivered this result in the real world.

This book uses World's Best Practice as the standard or benchmark against which all Defence Weapons Acquisition programs must be measured.

Why?

Because this standard has already been achieved by others and is therefore achievable in Defence Weapons Acquisition programs with effort.

Why is a standard or benchmark needed?

Simply because without a standard any result can be passed off as good - despite the actual result being bad.

Meaning 1: 'Where there is no standard - there is no variation'.

Or more correctly, you can't measure any variation - because there is no standard to measure against.

Or we can say. 'Where there is no standard - then every result (good or bad) is completely acceptable.'

Meaning 2: 'If you can't measure it, then you can't improve it'.

While action can be taken to improve an activity - we can't know if it is an actual improvement - unless some benchmark, standard or target exists to show the relativity of that change.

Is it better? Worse? We can't know because we don't know the prior baseline state before the change is made.

Implications for Defence:

As the reader will see in the following chapters, Defence obfuscates its standards and benchmarks in Weapons Acquisition Programs such that it is almost impossible to gauge what is a good result and what is a bad result.

That is, Defence applies a weak or very low standard, or baseline, to Weapons Acquisition Programs.

I have laid out this case in the following chapters so the reader can decide this for themselves.

In Part 2 - How to Drain the Defence Swamp - implementing World's Best Practice is the only acceptable Standard.

What is World's Best Practice?

Let me start with a simple metaphor.

The Olympic Games - 100 Metres in Athletics. The ultimate in World's Best Practice is the 100 metres Gold Medallist in Athletics. Or perhaps we could use the three medallists (Gold, Silver, Bronze) as defining World's Best Practice.

But that extreme level of excellence and performance is not achievable by most. A more practical World's Best Practice level would be all 40 people who reach the 100 Metre semi-finals. Or perhaps we could use the Olympic 100 Metre Qualifying Standard of around 100 people.

So, a *reasonable* World's Best Practice standard would be this: 40 to100 people who have already actually delivered this World's best level of performance, which can therefore reasonably be achieved by any other serious athlete prepared to put in the same hard work and effort.

And so it is in Industry and Defence.

World's Best Practice and Defence

If 40 to 100 of other firms around the world can deliver a World's Best Practice level of capability and performance, then any Defence firm, or Defence itself, can have no excuse why they can't reasonably achieve the same level of World's Best Practice if they are seriously prepared to put in the same level of hard work and effort.

Weapons Acquisition is no different to any other difficult product - and therefore Defence is not entitled to assume they should somehow be held to a much lower standard.

Defence must be held to a reasonable and practical World's Best Practice Standard that has already been achieved by others.

Defence has no excuse that can justify otherwise.

An early Introduction to World's Best Practice 2020 (WBP2020)

Beginning in Chapter Four a new methodology and capability will begin to emerge which I will term WBP2020. (World's Best Practice 2020). In Part 2, WBP2020 is proposed as the best solution to drive weapons acquisition reform as the best solution to the current Defence Swamp.

Prelude to key concepts in this book - (a quick summary to introduce key points right up front):

My own personal work experience mirrors the two halves of this book.

My work-life began in Product Development working for 15 years in heavy Industry using much the same reactive thinking and methods described in Part One - the Defence Swamp. I too was incompetent, but unaware of it.

That is, until I joined the Toyota Group, where reactive thinking and methods are abhorred. Joining the Toyota Group demanded a complete and radical shift in my thinking from reactive to proactive thinking and methods, along the lines of that described in Part Two - the solution to the Defence Swamp.

For the past 30 years I have used the thinking and methods described in Part Two to deliver far better outcomes both for myself (inside the Toyota Group), and for other commercial world firms after I left the Toyota Group. So, I know this stuff works well when applied correctly.

It also currently works well for astute firms in the commercial world.

It can work well in Defence Weapons Acquisitions - if Defence is prepared to use it.

The Reasons You Should Read This Book:

The Bad Bits:

We used to be much better than this, but now we have advanced to the same kind of mess as everyone else.

Things in Australia's Defence Weapons Systems Acquisition are really quite 'shit' – among the worst in the western world.

But it wasn't always this bad. There have been past Weapons Acquisition programs that show clearly that Defence used to be more competent at Weapons Acquisition than they are now. There have even been some more recent individual programs that have delivered success. The problem is they are all too few, and all too far between.

In fact, things are getting worse, not better, despite Defence's claims to the contrary, which assume that you have forgotten about the past.

It's much the same with the individual firms supplying Defence. We used to be much better than now. Today there are only a very small number of very good, highly competent firms in the Defence supply chain. The problem is they are too few, and too far between, to have any impact upon the total end-to-end Weapons Acquisition System outcomes.

The Good Bits:

What is proposed has already worked with some Defence projects – C-17 Aircraft and Astute Submarine

What is proposed works well for Toyota and has worked well for me personally as a GM/CEO.

What is proposed will most definitely fix the current 'broken' Defence Weapons Acquisition failures.

Commercial world firms have used these new methods to dramatically reduce the cost of their products to such an extent they have been able to avoid off-shoring their products and the Jobs of their people to China.

I have received so much support from so many people from inside Defence and Industry, including senior leaders who have been willing to talk openly about the past failures. They are prepared to talk to me because they are seriously "shit-to-bits" (*their words*) with having to labour inside the current completely 'broken' Weapons Acquisition systems, and which prevents them from delivering best quality and best outcomes to the Warfighter.

These people are true patriots, they truly care about Australia, they care about the obscene waste of Taxpayer's money, they care about the dysfunction and incompetence in the current methods, and they truly care about delivering the best possible Weapons outcomes battle-ready into the hands of the Warfighter.

An Introduction to the Chapters:

Part One opens the case for reform by laying out the failings of the current methodology.

While the flow of early chapters tells its own story, two chapters need a special introduction. These are two ways that what are perceived to be core failings can be viewed.

Chapter 5: The Incompetence View

This chapter takes the view that the persistent failures of Defence Weapons Acquisition programs is bad because using incompetent methods and incompetent thinking creates organisational incompetence.

This organisational incompetence in turn feeds the on-going failures in Weapons Acquisition programs.

This circular Organisational incompetence then means that Defence is unable to break out of this endless cycle.

Chapter 6: The Wall Street View

Wall Street: The single-minded focus where *share price* of any firm is the *only* thing that matters.

This chapter discusses an alternate view whereby cost blowouts and long-time delays are not bad, but, in fact, are good for the profitability and share price of the individual Defence Firms involved.

But, of course what's good for the individual Defence Firms is not in the best interests of the Taxpayer.

This may sound like all Defence Prime contractors are villains who are in it solely to make obscene money.

Perhaps they are, but we do have to ask the question of whether this is all their fault, or is it the fault of the Defence bureaucracy who continue to allow reactive and basket-case programs to happen time after time?

For me, this is simply the incompetence view observed from another angle.

The incompetence view simply allows the Wall Street view to happen.

But I do accept that there is an alternate view which thinks the Wall Street view is itself a valid business model.

Either way there are some obscene amounts of taxpayer money wasted in Defence Weapons Acquisitions.

This is a question you as the reader will need to consider for yourself as the story in Part One unfolds.

Chapters 8 to 11 then explain precisely why Defence and Politicians can't fix it, and therefore why the taxpayer will continue to pay for the obscene cost blowouts and the Warfighter will continue to pay for the delayed capability.

Part Two describes a journey to a whole different possibility.

While I have sometimes used the term 'Bold and Daring Reform' in this book it is not really all that bold and daring to those for whom the thinking and methods described in chapters 15 to 19 just makes total sense.

But for those stuck in the status quo of reactive thinking and methods (Defence, the Defence Bureaucracy, and our Politicians) it is a very bold and daring reform proposal that is perhaps far beyond their ability to deliver.

So, my use of 'bold and daring' is more a challenge to the way of thinking than about a revolutionary outcome.

The methods I have described in Chapters 15 to 19 are really nothing completely new. They are only completely new to Defence. They have long been used in earlier forms by the most astute businesses around the world.

So, while Defence continues in 2020 to use their existing failed methods as described in Part One, the approach described in Part Two is long proven to deliver better outcomes in practice to the commercial world.

While there are no new weapons programs I can demonstrate as an example - there are a lot of remediation examples of Defence Weapons already using this new thinking to deliver better outcomes around the world.

A Defence Example - McDonnell Douglas/Boeing C-17 (from Chapter 16 - Example B)

An earlier version of the methodology described was used in a partial remediation program during the development of the C-17 heavy-lift aircraft. The subsequent remediation fix road map when implemented has resulted in C-17 now being regarded as one of the most successful Airforce Programs when in operational use in the hands of the Warfighter.

- **Part Reduction** = 46,135
- **Operating Steps Reduction** = 273,008
- **Time (hrs) Reduction** = 5,550
- **$ Saving per Aircraft** = $1,371,001

Raytheon: 18 Year Problem... Solved In 4 Months - 50% Reduction in Weight - 72% Reduction in Cost
Astute Submarine: Cost Reduction Program on Boats 4-6 (delivered targeted Cost Reduction amount)

What's sad about these remediation examples is that they show that while Defence around the world does not use these new methods to develop their new weapons systems or equipment in the first place, they are only too ready to use these new methods to fix the 'broken' weapon or equipment, but only after it is 'broken'.

The second thing that is sad about these examples is there are no Australian examples of new weapons systems or equipment, and no remediation examples because our Defence does not use any better methods to remediate 'broken' equipment than to use the very same methods which broke it in the first place.

For you, the reader, it is now time to begin a journey into the gloom, despair, and darkness of the first 13 chapters.

But you should not despair - the first faint spark of a new light begins in chapter 4, which will become a little brighter still in Chapter 7.

But it is not until Chapters 14 to 21 in Part Two before you will see how to farewell that gloom and darkness forever - and look forward towards a far better, much brighter future.

Notes:

1. Defence uses the term 'weapons systems' instead of the term product, or equipment, because it is not simply about a single piece of equipment - but the creation of a fully capable battle-ready Weapon System. The terms weapons systems (plural) and weapon systems (singular) are used throughout this book.
2. Defence uses the term 'capability' in the hands of the Warfighter to describe the required weapon. First comes the capability, followed by the specific equipment that will deliver that explicit capability.
3. All A$ Currency conversions use an A$ to $USD conversion of $0.75 (past 5-year average exchange rate)
4. For those interested in Costing of Defence Programs- ASPI (*Australian Strategic Policy Institute*) has released a new *Cost of Defence Public Database*. ASPI is one of the more reliable places to find verifiable numbers for Defence Programs, although that does not stop them from occasionally suffering from the same 'Defence disease' that will become evident to you as the early chapters and case studies unfold. This *Cost of Defence Public Database* has no impact upon the numbers I have used throughout this book but is a good resource for any reader interested in further exploring the 'unaffordability' of Defence Acquisition Programs.

I

THE DEFENCE SWAMP - THE VITAL NEED FOR REFORM

SWAMP: Traditional meaning: a tract of wet, spongy land, that is <u>wholly unfit</u> for cultivation.

SWAMP: Current usage: Government Bureaucracy of dysfunction, waste & incompetence, wholly unfit for purpose, and wholly unfit for office.

Note: "Drain the Swamp" has most recently been alluded to by Donald Trump during, and since his campaign to become President of the United States. In his inaugural address Trump said: "For too long, a small group in our nation's capital has reaped the rewards of government, while the people have borne the cost."

In Trump's terms he has used this drain-the-swamp metaphor referring to reducing the seemingly corrupt and out-of-control bureaucracy across many areas of the United States government. He describes it as tacitly understood to be a visual analogy for what was occurring in Washington: a site of stagnation, stench, and rot.

Translated into political imagery, the Swamp has been described as the source of dysfunction, waste, and deceit.[1]

My Meaning:

I will limit my use of the term 'swamp' to describe the abysmal results from Australia's dysfunctional Weapons Systems Acquisition programs and the methodology that has been, and is still being used today, to produce them.

I will describe the 'broken system' system used by Defence to develop Weapons Systems - with the cost blowout penalties borne by the Australian taxpayer and the capability short-comings borne by the Warfighter from these 'broken systems'.

The extreme incompetence demonstrated in the following pages is, as always, a failure of Leadership alone, not a failure of those who vainly try, often to no avail, as best they can, to work within the 'broken systems' that are Defence Weapons Acquisition.

While Defence is an institution that Australians continue to have faith and respect in, *it nonetheless has not been immune from decay, cronyism, and dysfunctional, sclerotic, slow changing bureaucracy.* [2]

The Swamp 2020:

Given the severe economic impact from Covid-19 the entire Nation's future economic strategy is now in question. Our Government can no longer allow the ineptitude and blatant waste of taxpayer's hard-earned money to continue in the normal way that is applied today in the Defence Weapons Acquisition swamp.

Now more than ever we need to 'drain the swamp'. Australia needs a new way forward.

How to Drain the Swamp:

1. Create an Organisational Bureaucracy of a new World's Best Practice - competent, functional, highly effective, wholly fit for purpose, and wholly fit for office.
2. The Australian People must demand a much higher standard of competency and performance from all of those who serve us if we are to take our Nation forward into the economic and social future we now face.
3. This solution to the Defence Swamp must deliver this new higher competency and performance requirement from each and every Weapons System Acquisition program.
4. Nothing short of total reform of Weapons Acquisition will be acceptable.

5. Nothing short of complete disruption of the Defence Swamp will be acceptable.

A First Narrative:

Scene #1: Far North Queensland after Cyclone Yasi tore through with utter devastation.

Problem: Defence had zero ships available to provide aid in the aftermath.

Scene #2: A week later at a major Defence conference in Canberra.

Defence Minister Stephen Smith opens the conference by giving Defence a right old bollixing. In crude vernacular he ripped Defence a new arsehole. Smith finishes, leaves and returns to Parliament House. Conference continues.

The Chief of Defence does not respond in any way to the Minister at all - presents his talk entirely as pre-prepared. The CEO of Defence Materiel follows, does not respond to the Minister - presents his talk entirely as pre-prepared.

By morning tea, it was: "Smith? Was Smith here? Smith who?"

In hindsight, it was a very big red flag about the Defence Way.

If Defence can so blatantly ignore a Minister of the Crown, their own boss, then what hope for us mere mortals?

---

1. *Jeff Groom, The Military Swamp: I Got out, but Left a Broken and Stagnant Ecosystem Behind, questia.com, Dec 2018*
2. Jeff Groom, ibid

1

THE SWAMP

Basic Principle: The whole point of Defence acquiring Weapons Systems Programs and Equipment is so that our Warfighters can successfully prosecute any war.

Therefore, anything at all which impedes in any way our Warfighters' ability to successfully prosecute a war must be deemed a complete failure of our Defence Weapons Systems Acquisition programs and methods.

Note: there is no second place in the prosecution of war - you either win, or you lose.

There is no 'honourable' draw.

To be able to successfully prosecute war our Warfighters must be given the right capability to win every time.

If the equipment and capability they need is not immediately available to them - when they need it - with the reliability they need - our Warfighters, and our country, will lose.

Therefore, any failure in Weapons System Acquisition is a complete and utter failure to the Warfighter.

Note: Our Warfighters are unable to prosecute any war until the equipment is fully tested and battle ready in the hands of our Warfighters. Equipment is not ready to prosecute a war when some politician ceremonially hands it over. (Often it takes anywhere from 1 to 3 years for equipment to be tested and fully battle-ready).

Definition of 'Prosecute': to continue with (a course of action) with a view to its completion, especially a war.

e.g. 'to pursue the government's ability to prosecute the war.'

Prosecute Failure Modes: Anything which removes, or delays, Warfighter access to the required equipment precisely when they need it.

Government Policy:

- Political focus to other areas
- Delays to Defence White Papers
- Delays in actioning White Papers
- Lack of Budget to support White Papers

Defence Policy:

- Defence distractions away from the core purpose
- Defence Politics
- Delays in implementing Policy/White Papers
- Weak or non-existent Targets and Hurdles for Programs
- Poor Program Risk control
- Poor Program Management systems
- Lack of control over Defence Supply Chain
- Design & Development Programs

By Defence and/or the Defence Supply Chain

- Reactive thinking and methods
- Weak or non-existent Targets and hurdles
- Poor internal Program Risk control
- Poor internal Program Management systems
- Lack of control over their tier two Supply Chain

While this book concentrates exclusively upon Weapons Acquisition failure modes, in doing so it also describes much broader and deeper problems that portray a whole of Defence Swamp that is 'wholly unfit for purpose'.

Many of the problems which follow in Part One have their roots in the failure modes listed above.

These are failures of leadership, not failures of execution by those working within Defence's 'broken' systems.

These failures of leadership directly contribute to the Organisational Incompetence view of Defence.

These failures of leadership portray a Defence Swamp. A government bureaucracy of dysfunction, waste & incompetence, <u>wholly unfit</u> for purpose, <u>wholly unfit</u> for office.

> *"If a sufficient number of management layers are superimposed on each other, it can be assured that disaster is not left to chance."*
>
> *~ Augustine's Law Number XXVI*

.

2

ALTERNATE REALITIES

In Defence the use of alternate realities is the norm.

In a factual reality, any fact remains the same even if we view it from four different directions. In Defence's alternate realities the same fact can be interpreted in as many different ways as there are perspectives.

This is extremely disconcerting, and very confusing for all non-Defence people. And will remain so unless we can understand why these alternate realities exist in Defence.

I had a career spanning 23 Years working across the Toyota Group of companies. In Toyota Facts are facts, or nearly so. If I spend $11.36 that is a definite fact. But if I buy that from Japan that $11.36 can vary slightly because we do not recalculate the exchange rate every minute of every day, or even weekly, or monthly.

So, these sorts of facts are known in Toyota as 'near facts'. That is, they are 97%, 98%, or 99% accurate, based on an original fact. So, a fact is a fact. And a 'near fact' is, within a very high probability of accuracy, a fact.

In Toyota there was a third type of fact known as a 'bullshit fact.' For example, to engineers these were cynically and colloquially known as 'marketing facts'. Marketing facts exist where we can think of a number, any number, and repeat it endlessly until it becomes a fact. But is still a 'bullshit fact'. It has no basis in reality.

So now we have facts, 'near facts' and 'bullshit facts'.

When I joined Defence, I had to quickly learn that there exists another type of fact known as 'alternate facts'.

Confused? So was I.

Coming from a highly ethical environment like Toyota, 'alternate facts' seemed like a highly unethical way of doing business. But Defence is not a business. Defence is a government bureaucracy that works to Bureaucracy's rules. The rules of the commercial world do not apply.

> *"Truth is a matter of perspective. There will be my truth, and your truth, as universal truth does not exist."*
>
> *~GEORGE ORWELL*

Let me give you a practical example:

In Toyota we would create a waterfall chart to describe the cost difference between an old car model and a new car model. This chart would show all the negative and positive cost movements between the old and new costs.

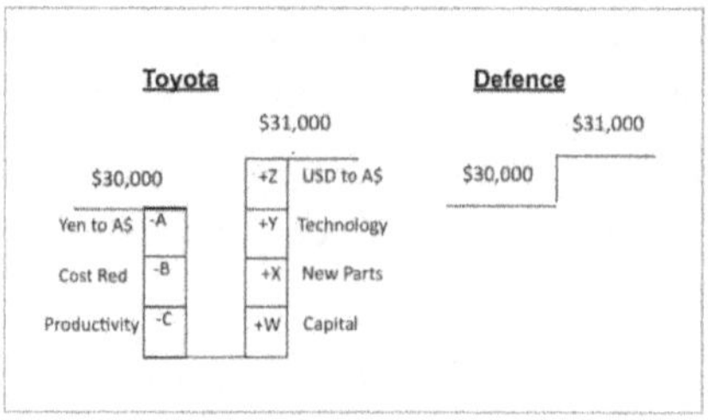

Figure 1: Toyota Waterfall Chart for Cost

In the Toyota model, every cost movement is shown as a fixed 'fact'. So, if anything changes it is obvious which factor has caused the change. But in the Defence model no details are provided, so it is easy for example to say that the cost increased to $32,000 because of new Technology. It is impossible to argue against this alternate fact because we were never told whether new Technology was included in the original $31,000.

This is precisely how 'fluid', 'muddy', or 'alternate' facts are created. Every number can be made to mean something different depending upon the circumstances needed to explain it.

If you think this is obfuscation of perspective, then you are right.

In Toyota using this waterfall model results in an ever-improving human ability to predict and forecast cost movements. In Defence there is never any improvement because there is nothing to learn from such 'fluid' numbers.

So why does obfuscation through these alternate realities exist in Defence?

Clearly, it is a rewarding behaviour of great benefit to Defence or they would not do it. Primarily they are used to protect and preserve the 'sacred cows' in Defence Weapons Acquisitions.

But if I put my Toyota hat on, this is simply to hide the incompetence of leadership.

Chapter Five will pick up and expand on this theme.

3

CASE STUDIES

(COLLINS - AWD - HCF - JSF - FS THE 'FRENCH FOLLY' - ALL EXAMPLES OF FAILURES)

Note: these case studies are not meant to be an exhaustive analysis of the programs. Rather they are used as simple primers to explain a point of view about the results of problems in Defence Weapons Acquisition programs.

If the reader wants more exhaustive analysis of any of these programs there are many books, reports and studies written about them all here in Australia by ANAO (Australia National Audit Office) or by ASPI (Australian Strategic Policy Institute). In the USA, the GAO (Government Accountability Office) and DOTE (Director Operation Test & Evaluation) or DAU (Defence Acquisition University) have produced a vast number of reports, studies and analysis for US Defence Weapons Acquisition programs.

These case studies will be like no other case study you have ever read.

Normally a case study represents a series of factual statements that took place. But this is Defence.

Defence uses alternate realities and alternate facts. Alternate realities are entirely 'normal.' In simple terms there are many sides to every story. So, something that I say is a failure, others may call a massive success, and both realities co-exist, and are both are seemingly valid. It all depends on your perspective and how you are measuring the results.

But measuring results is also another new set of alternate realities. Whatever results are required to suit the circumstances can be delivered in a series of alternate realities using alternate facts.

If you are confused, that is entirely the aim.

It is impossible to pin down anything or anyone precisely. Everything exists in a 'muddy' pond of information. Everything can be explained depending upon how one views the 'muddy' information.

Obfuscation of the facts is the aim.

The following five case studies illustrate this point.

Each presents two alternate realities: A Defence reality, followed by my reality (with my Toyota thinking hat on).

It is for you the reader to decide which alternate reality best describes the success or failure in each case study.

The numbers I have used in each case study are from my notes taken during Defence meetings and conferences, or from reputable sources like ANAO or ASPI. None are from general media as their numbers are not verifiable.

I have sought only to use real facts or 'near facts' of a very high probability of being correct.

No 'bullshit facts' or other fanciful numbers have been used.

It is important for the reader to understand that even if every number in this book can be proven beyond any shadow of doubt to be 100% perfectly accurate, it will not change the underlying belief in Defence about the success or failure of each of these case studies.

In Defence, opinions are far more important than facts.

> *"Everything we hear is an opinion, not a fact. Everything we see is a perspective, not the truth."*
>
> ~ MARCUS AURELIUS

Each case study will be followed by questions for you the reader.

My job in this book is to lay out the facts as I see them. Then paint alternate realities as Defence sees them. And then let the readers decide for themselves among these alternate realities where the best actual truth lies.

Reader's Analysis: This section will explain and lay out a direct comparison between the views expressed in the alternate realities in that case study.

Reader's Choice: You must then choose for yourselves which of these alternate realities is where the best truth lies, and which of the alternate realities best describes program success or failure.

> *"A foolish consistency is the hobgoblin of little minds, adored by little statesmen and philosophers and divines"*
>
> ~ *RALPH WALDO EMERSON, SELF-RELIANCE*

A 'foolish consistency' is one of Emerson's recurrent themes: the need to avoid conformity and false consistency whenever one set of data are made to contradict another, simply for the sake of a recurrent consistency.

A. Collins Class Submarine

For a simple overview of Collins go to Wikipedia: *Collins-class submarine.*[1]

The ***Collins* class** consists of six Australian built diesel electric submarines constructed from 1990 to 2003.

Collins Alternate Reality A: The Defence Way

Planning for a new submarine to replace the *Oberon*-class submarines began in the early 1980s, with a winning design announced in 1987 for an enlarged version of Swedish shipbuilder Kockums *Västergötland* class which were constructed between 1990 and 2003 in South Australia by the Australian Submarine Corporation (ASC).

These boats were the first submarines to be constructed in Australia, prompting widespread improvements in Australian industry and delivering a sovereign (Australian controlled) sustainment/maintenance capability.

The submarines were beset by problems in their early lives, including the failure of a major contractor to deliver a combat system that was fit for purpose. Problems with the system of maintenance and upkeep for the submarines also contributed to sub-par availability. Defence invested a great deal of effort into analysing and successfully overcoming these challenges.

The decision to build submarines in Australia for the first time was bold and courageous. Defence had learned how vital it was for having a reliably available capability that we understood the submarines deeply and could draw on local industry and supply chains to keep them properly maintained and upgraded with the latest technological developments when necessary. The decision to build a unique submarine design was driven by the fact that there was nothing in the market able to meet our needs, which were a function of our geography. We made mistakes, some of them expensive, but Collins was, in fact, a stunning national success story that has never been properly acknowledged as such. They came in very close to the global benchmark cost for a submarine of its size (based on $/tonne displacement) and had a very high Australian Industry content. This target was deliberate and has since served us very well during repair, maintenance and upgrades undertaken since. The construction quality was world-class.

Many of the problems encountered in those early years arose because we trusted overseas specialists on things where they had more experience than we had in Australia. Eventually it became clear that they could not fix the problems we were experiencing, which we then drew on our own engineers and scientists to solve. We also forged a much closer partnership with the US Navy. In addition to fixing the design problems and flawed sustainment program we had been sold, Defence had to sort out an incompetent firm doing the sustainment industrial activity. These activities all required injection of additional funding, but the efforts have served us very well. Today, we have a submarine giving us global benchmark availability at benchmark cost with 90% Australian industry and supply chain involvement. All this experience has been examined in detail to learn the lessons which are highly valued and being applied in the Future Submarine Program.

Collins Result: Acquisition SUCCESS.

Collins Alternate Reality B: The Systems Way:

This is perhaps the most familiar (and notorious) of Defence weapons programs for most Australians.

The initial operational plan for 6 submarines.

- 3 to be fully operational in the water at all times
- 1 to be 50% available for operations, plus 50% available for crew training

- 1 in short cycle docking (maintenance and sustainment)
- 1 in long cycle docking (sustainment, repairs and upgrading)

To ask the Dr Phil question: How did that work out for you Australia?

- For some periods of time we had no submarines fully operational in the water.
- For most of the time we have had only 1 or sometimes 2 submarines operational in the water.
- Poor availability of submarines severely crippled crew training and especially new crew training.

It is only now, 25 years later, that the original operating plans are close to being delivered.

Note: When it is fully operational in the water, the Collins was the best submarine in the world on its day. Our problem is simply that we could not keep them operational in the water on a continuous basis.

Note: the Collins class was due to be retired from 2025 - but dramatic delays in the future Submarine Program (see 'French Folly') will mean the Collins service life will need to be extended for up to 15 years.

Therefore, the Australian taxpayer will pay twice at the same time. Once for the new submarine and also, at the same time, for the extensive updates necessary to Collins to extend its life instead of retiring them as planned.

Key Points:

1. The six Collins boats suffered a quite severe cost blowout A$3.9b to A$6.2b, and schedule delays of 18-41 months during design and build.
2. Collins sustainment (ongoing annual maintenance) costs are around <u>double</u> what they should be.
3. Collins will need a future refit costing up to $15b to extend its service life to 2040 and beyond.
4. But the critical failure in Collins has been that our Warfighters would not have been able to successfully prosecute a war, because for much of the past 25 years the submarines were simply not available for use.

Note: if anyone disputes this, then simply ask them how we could prosecute a war with one (or NO) submarines?

Remember: It is the ability to prosecute war that matters - not the weapons program itself.

What the reader needs to take-away from this is that the physical Collins Submarine itself is not the problem.

A 'broken' design & development system will <u>always</u> deliver a 'broken' product (submarine) as the outcome.

The Collins Submarine with all its faults and problems is simply an absolutely perfect outcome from the completely 'broken' system used to design and develop it.

We designed the submarine to be 'broken' - and it was.

And then we express surprise and anger that it is broken!!

It is the current methodology used to design, develop and acquire Defence Weapons Systems that is the problem.

But to the Warfighter - it matters not the cause of the problem - only that these problems prevent the Warfighter from successfully prosecuting war.

Did this Weapons Acquisition Program meet the World's Best Practice standard or benchmark?

Would this submarine weapons system have allowed us to successfully prosecute war if we had to?

No. Only with a very low or very weak standard could Collins acquisition be regarded as successful.

COLLINS Result: Acquisition FAIL.

Reader's Analysis:

As you can see the Defence Alternate Way and the Systems Alternate Way describe completely different program outcomes. One success and one failure.

We can now see that the facts and the conclusions of success or failure can be seen vastly differently.

Was the Collins Submarine in the hands of the Warfighter 'wholly fit for purpose'?

Was the methodology used to design and develop the Collins weapon acquisition 'wholly fit for purpose'?

Reader's Choice:

Now you must choose for yourself, which of these alternate realities is where the best truth lies, and which of the alternate realities best describes program success or failure.

Which alternate reality best described a 'wholly (fit) or (unfit) for purpose' Weapons Acquisition program?

Alternate Defence Reality A? or Alternate Systems Reality B?

B. Air Warfare Destroyer (AWD)

For a simple overview of AWD go to Wikipedia: *Hobart-class destroyer.*[2]

What we now know as the three-ship set of *Hobart* class Destroyers began life as a 3-way contest between German, Spanish and American designs. Despite the USA evolved Arleigh Burke being the preferred design in 2005, by 2006 the Spanish ships were considered a less-risky design as the Spanish vessels had already been built and were operational. The German option was rejected early in the competition.

In 2007 the *Spanish Navantia* design was selected over the USA Arleigh Burke on the assumption that, as it was an existing design, it would be cheaper, quicker, and less risky to build (*or so they thought*).

The *Navantia* derivatives were predicted to be in service *four years earlier* than the American-designed ships and would cost A$1 billion less to build. (*Ha-ha!*)

AWD Alternate Reality A: The Defence Way

The Air Warfare Destroyers would be the most complex warships Australia had ever built, with significant strategic importance for continuation of Australia's Defence Industry. Several key decisions framed the program.

The first was to select the air warfare weapon system, the heart of the ship's capability. The US Navy's Aegis system was the obvious choice. It was well-proven, with over 3000 missile firings and there were around 100 systems in service with five navies (Japan, South Korea, Spain and Norway in addition to the USA). It also provided Australia with the required interoperability with the US, where it be in service for the next 30 years or so and acquisition and support costs were well known and understood. Upgrades to provide ballistic missile defence (BMD) and the emerging cooperative engagement capability (CEC) were on the way and both were attractive to Australia.

Two alternative systems were available, the German-Dutch APAR based system and the British-French PAAMS system. Both were in development and therefore unproven, with largely unknown or unproven capability, costs and risks. They would only ever be in service in much smaller numbers than Aegis.

Next, Defence selected the shipbuilder. The Government's own warship builder, ASC Pty Ltd in Adelaide, won in stiff competition against several other local companies.

Finally, after a rigorous competitive process involving exhaustive analysis of the German, Spanish and US options, Defence decided that the risks overall would be managed and mitigated most effectively by choosing a proven design that required minimal change. The Germans had no experience with Aegis and the US ship existed only 'on the drawing board'. Since it was already in service and embodying the Aegis system, the Spanish design was selected.

After construction of new shipyard facilities in Adelaide and some early teething problems, the AWD program delivered all three ships to world class quality standards in a stunning success for Australian industry and Navy capability.

Note: ASPI even commissioned a book laying out what a stunning success this program was!

AWD Result: Acquisition SUCCESS.

Alternate Reality B: The Systems Way

To ask the Dr Phil question again: How did that work out for you Australia?

A simple Cost Table:

- The rough *Market Price* of each *off-the-shelf* AWD from Navantia was around, or just below A$1b
- Remove Spanish weapons system, add US weapons system - market price of ~A$1.2b each

- Now add an extra ~A$400m to build in Australia - base budget A$1.6b each
- Add A$200m for contingencies - full local build budget A$1.8b each
- First stuff-up - budget revised from A$1.8b to A$2.4b each
- Second stuff up - budget revised to A$3.0b each
- Final cost of fully battle-ready ships (late 2021) will be around A$3.4b each

For the full three ship set, the costs went from around A$3b market price to A$4.8b for the base budget, or A$5.4b for the budget with contingencies, which then blew out to ~A$10.2b by the completion of three battle ready ships.

Note: Our Warfighters are unable to prosecute a war until the ships are fully Battle ready which (for a program started in 2007) will not be until 2021 at the earliest. (14 years to finally provide the capability to our Warfighters).

Final result: A$1.6b cost blowout each - a little less than A$5b cost blowout for 3 ships. This is also A$6.6b more than a reasonable global Market price for these ship types.

Note: the US is planning its Arleigh Burke replacements at around ~US$930m each (roughly A$1.2b each).

Oh, and by the way the AWD's were delivered more than 3 years late! (according to the original schedule).

So, let me get this straight... we chose the Spanish design because it was already operational in the water. Therefore, we only had to copy the existing manufacturing of most of the ship (plus any redesigns to fit the US weapons systems). So, an 'easy' design that was supposed to save $1b and be in the water 4 years earlier ended up costing A$5b more than planned and was in the water no earlier than the US better design would have been.

Note for the reader: It will be difficult to trace these numbers because Defence re-bases the budget constantly.

i.e. after the first stuff-up the A$1.8b number is deleted and only the A$2.4b number is available. Ditto after the 2nd stuff-up, the A$2.4b number disappeared to be replaced by A$3.0. So, the current program budget (2017) shows the AWD program is only around A$100m over budget (such obfuscation also applies to program timelines).

Additionally, no-one mentions either that these three ships are replacing three previous AWDs, the last of which retired on 19 October 2001, making for a trifling 20-year capability gap for our Warfighters.

Key Points:

1. AWD suffered very severe cost blowouts and long schedule delays during design and build.
2. AWD expected sustainment (ongoing maintenance) costs will be much higher than they should be.
3. But the critical failure mode in the AWD Weapons Systems acquisition process has been for the Taxpayer.

For the final cost of around A$10b the Taxpayer could have bought for our Warfighters:

1. Ten ships off-the-shelf from Navantia (with Spanish weapons systems).
2. Eight ships from Navantia with US weapons systems.
3. Six ships built in Australia at the original budget level - and on time.
4. Three ships at a blown-out cost of A$10b and 3 years late.

These are now considered the most expensive naval ships size-for-size in any navy in the world.

Which do you think the Taxpayer would have preferred?

Which do you think the Warfighter would have preferred?

Which did the current Defence Weapons Systems Acquisition process deliver?

Did this Weapons Acquisition Program meet the World's Best Practice Standard or Benchmark?

No. Only with a very low or very weak standard could this AWD Acquisition be regarded as successful.

AWD Result: Acquisition FAIL.

Reader's Analysis:

As we can see the Defence Alternate Way and the Systems Alternate Way describe completely different program outcomes. One success and one failure.

We can now see that the facts and the conclusions of success or failure can be seen vastly differently.

Was the methodology used to Design and Develop the AWD Weapon Acquisition 'wholly fit for purpose'?

Reader's Choice:

Now you must choose for yourself which of these alternate realities is where the best truth lies, and which of the alternate realities best describes program success or failure.

Which alternate reality best described a 'wholly (fit) or (unfit) for purpose' Weapons Acquisition program?

Alternate Defence Reality A? or Alternate SYSTEMS Reality B?

C. Hunter Class Frigates (HCF)

The *Hunter*-class frigate is a future class of frigates to replace the (1980's) Anzac-class Frigates. Navantia, Fincantieri, and BAE Systems were preselected in 2016 as contenders to design & build the ships.

In June 2018 BAE Systems Type 26 ship was chosen as the basis of HCF.

HCF Alternate Reality A: The Defence Way

The new Hunter Class frigates are a substantial component of plans to totally modernise Australia's anti-submarine warfare capability. Other components of this major program and massive investment include the Attack Class submarines, the MQ4C Triton very-long endurance maritime surveillance pilotless aircraft, new P8 Poseidon long range maritime patrol aircraft and the MH60R Seahawk helicopters that will operate from the frigates.

The Type-26 is the latest in a long line of sophisticated warships designed in Great Britain that are optimised for ASW. It will be equipped with the most advanced sonar suite in the world today. Australia's version of the ship will also be fitted with the world's leading phased array radars which were designed and are produced by Australian company CEA Technologies in Canberra. These sophisticated radars are proven in service at sea in the Anzac class frigates that the Hunter class ships will replace. The new frigates will also be fitted with the Aegis Weapons System common to the Hobart Class air warfare destroyers, the world's most advanced air defence system, combined with the SAAB Australian interface, making these ships very potent additional war-fighting capability for Australia's naval fleet.

The Hunter class acquisition is a significant element of the Government's commitment to a continuous naval shipbuilding program. Vital to the success of that policy is an aggressive schedule and requirement for maximum Australian industry involvement with the new ships. These plans are firmly on track, with steel manufactured by Australian steel producers to be cut later in 2020 for qualification modules in preparation for the first ship. This work will be done at ASC's shipyards in Osborne, just outside Adelaide, where the very successful Air Warfare Destroyer program has recently concluded.

HCF Prediction: High probability of Acquisition Success.

HCF Alternate Reality B: The Systems Way

While HCF is a slightly higher tonnage, dimensionally HCF and AWD are very similar - HCF being 2 metres longer and 2 metres wider (150m x 21m v 148m x 19m).

It is instructive therefore to understand the cost budget for HCF with relationship to AWD actual costs.

The HCF is planned to be a A$35b program.

Allowing for approximately A$7b for infrastructure and other costs - we get approximately $28b for a build of nine ships.

This equates roughly to $3.1b per ship.

This is precisely the same number as the current end cost of each AWD ship!

We can infer from this that Defence has 'normalised' the dysfunction of the AWD ship development into the standard budget cost of the HCF ships. That is, HCF development will not be any better than the problem fraught AWD program. Costs cannot be reduced back to a more reasonable $2b (or less) per ship (extrapolating the original AWD budget of $1.6b - or $1.8b for the full budget with contingencies).

We can therefore infer that Defence is not looking to apply any better development methods for HCF than the dysfunctional development methods used on AWD.

Remember: USA Arleigh Burke replacements (154m x 20m) at around ~US930m each (roughly A$1.2b each)

What is the rough off-the-shelf Market price for a ship of this size and type in 2020?

A$1.2 to A1.4b?

Has Defence ever justified the difference between this and A$3.1b per ship?

What is the baseline cost they are working from?

Do they even have a baseline cost to measure all variations from?

A more reasonable $1.6 to $2.0b budget would require some type of reform of the acquisition methodology.

Therefore, we can assume no acquisition reform program is included in the HCF program.

Therefore, we can assume that HCF will deliver similar results to AWD.

Note: the definition of insanity: doing the same thing and expecting a different result.

Defence will argue that they have been through a first principles review which restructured DMO (Defence Materiel Organisation) into CASG (Capability Acquisition and Sustainment Group) and have redesigned the old ASC (Australian Submarine Corporation) into BAE's shipbuilding yard. (I will debunk this myth later in this book).

But as I stated back in the Introduction on page 3: (actually, first written in 2013 to the then Defence Minister)

"Unless actual root causes are addressed, no amount of additional oversight, extra regulation, rearranging of organisation boxes, creation of new offices, or changes to processes will help current or future programs."

To ask the Dr Phil question: How do you think this will work out for you Australia?

- HCF is based on the UK Type 26 ship with quite major Australian modifications.
- The first UK Type 26 is not due to be fully commissioned until

2027 - HCF design must be finished by 2023 - so all learnings from UK Type 26 will not flow back to HCF before our design is finished.
- HCF will use most of the same suppliers involved in the AWD program failures. What is now different?

Which cost budget do you think the Taxpayer would prefer?

Which outcome do you think the current Defence Weapons Systems Acquisition process will deliver?

Does this Weapons Acquisition program currently meet the World's Best Practice standard or benchmark?

No. Only with a very low or very weak standard can HFC Acquisition be predicted to be successful.

HCF Prediction: High probability of Acquisition FAIL.

Reader's Analysis:

As we can see the Defence Alternate Way - and the Systems Alternate Way - describe completely different program outcomes. One success and one failure.

We can now see that the facts and the conclusions of success or failure can be seen vastly differently.

Is the methodology being used to design and develop this HCF Weapon Acquisition 'wholly fit for purpose'?

Reader's Choice:

Now you must choose for yourself which of these alternate realities is where the best truth lies, and which of the alternate realities best describes program success or failure.

Which alternate reality best described a 'wholly (fit) or (unfit) for purpose' Weapons Acquisition program?

Alternate Defence Reality **A?** or Alternate Systems Reality **B?**

D. Joint Strike Fighter (JSF)

What we now know as JSF began life as a *Defense Advanced Research Projects Agency (DARPA)* project named **Common** *Affordable* **Lightweight Fighter (CALF)**

CALF evolved into a two-horse contest between Boeing and Lockheed Martin back in 1996.

In 2001 Lockheed Martin won the contest with an initial order intent of around 3500 planes with the simplest US Air Force pricing model originally to come in at US$42M unit flyaway cost each, including non-recurring costs, which plus other development costs equalled around $200b of total program cost.

For a brief overview of the JSF go to Wikipedia; Lockheed Martin F35 Lightning II.[3]

JSF Alternate Reality A: The Defence Way

With the approaching requirement to replace the F111 bombers that had been in service since the mid-1970s and which were becoming prohibitively costly to maintain, and the ageing fleet of F/A18 Hornet fighter-bombers, Australia decided to join the F-35 JSF development program with the US in 2002. The JSF was then and remains today the only 5th generation combat aircraft in existence or in prospect that would be available to Australia. This is one of the primary causes of its relatively higher cost and has also contributed to delays in development.

The JSF is a quantum leap in capability over previous generations of combat aircraft. It's a game-changer for air combat. This reality is not generally understood by the JSF's critics, who compare its performance against earlier generations of aircraft that can only be used in the 'old' way. The JSF's considerable capability will provide a very significant air combat capability which will also be further enhanced by the recent acquisition of the EA-18 Growler electronic warfare aircraft. These aircraft have much in common with Australia's Super Hornets that replaced the F-111's. *(which we only had to buy because the JSF was running so late!!)*

Delivery schedule delays and cost escalation have attracted widespread public attention over a very long period. This is not uncommon with innovative and complex new capabilities, especially combat aircraft. Australia had a similar experience with the revolutionary and potent F-111 'swing wing' bombers, *(yes they were so late that we had to lease 24 F-4*

Phantoms as a stop gap between the retiring Canberra Bombers and the F-111's. This behaviour by Defence is habitual!!) which went on, once problems were resolved by the US, to deliver to Australia a very valuable capability that remained unrivalled in our region right up until their retirement in 2010. Defence is very confident that the few remaining problems with the JSF are well on the way to being resolved *(few? - not if you believe DOTE!)*.

Australia's JSF acquisition program is on track and deliveries of aircraft to the RAAF are progressing well. The 20th of the planned purchase of 72 aircraft will be handed over this year. Transition to this new capability from the Hornets is proceeding smoothly.

Australian industry has also benefited from the JSF program's innovative global supply chain concept, winning significant work supplying high technology components.

JSF Result: Acquisition SUCCESS.

JSF Alternate Reality B: The Systems Way

This Case Study shows that the USA have even bigger problems with Weapons Systems Acquisitions.

I don't intend to go into any specific detail here. Many tens of thousands of pages of reviews and commentary have been written by hundreds of others about this project over the past 15 years. Little of it good. Most of it bad.

Suffice to say that as a Weapons Acquisition program it falls into the basket-case category. Endless program delays, endless cost blowouts, endless problems with the plane, endless software issues, and endless hand wringing on Capitol Hill have all come together as a supreme demonstration of all that is wrong with US Weapons Acquisition.

The only positive from this basket-case Weapons Acquisition program has been that the extreme level of angst and dissatisfaction with this program has generated a new awakening in the US Congress. This can best be described as 'there must NEVER be another program like JSF', or 'JSF must be the last ever program like this'.

Every time I read reports about JSF from the US GAO (Government Accountability Office) or DOTE (Director Operation Test & Evaluation) I am reminded of Arthur C. Clarke's 1951 science-fiction short story, Superiority.

Clarke's story depicts an arms race during an interstellar war.

Superiority (see Appendix C) tells the story of how a scientific and technically superior world bewitched by creating amazing new weapons is defeated by the *'inferior science'* of their enemy.

This story could have been written today as a metaphor for the F-35 Joint Strike Fighter Program, as well as other 'one-size-fits-all' or 'bewitching' high-cost weapon systems.

Almost 70 years ago, in an age of far lesser technology that could not have foreseen the 21st century, Clarke has written about the chase for bewitching technology that is the nemesis of the JSF program today.

Likewise, JSF constantly reminds me of Augustine's Laws (see Appendix B). In one aphorism, he plotted the exponential growth of unit cost for fighter aircraft since 1910 and extrapolated it to its absurd conclusion. Nearly three decades later, (1990's) he said, 'we are right on target'. By 2020, it must now be absurdly ahead of target.

Augustine's most cited law shows that defence budgets grow linearly but the unit cost of a new military aircraft grows exponentially:

> *"In the year 2054, the entire defence budget will purchase just one tactical aircraft. This aircraft will have to be shared by the Air Force and Navy 3½ days each per week except for leap years, when it will be made available to the Marines for the extra day."*

> *~ AUGUSTINE'S LAW NUMBER XVI*

Or perhaps this one:

> *"Aircraft flight in the 21st century will always be in a westerly direction, preferably supersonic, crossing time zones to provide the additional hours needed to fix the broken electronics."*

> *~ AUGUSTINE'S LAW NUMBER XLIV*

Or perhaps this one:

> *"Software is like entropy. It is difficult to grasp, weighs nothing, and obeys the Second Law of Thermodynamics: i.e., it always increases."*

> *~ AUGUSTINE'S LAW NUMBER XVII*

The prescience of Augustine is quite amazing. Almost 40 years later his laws today seem so real, so practical, so true. Even though he made them partly in jest.

Nemesis seems to always follow Hubris

But back to the Facts.

JSF began life as a DARPA project named: Common Affordable Lightweight Fighter (CALF). CALF evolved into a two-horse contest between Boeing and Lockheed Martin back in 1996.

In 2001 Lockheed Martin won the contest with an initial order intent of around 3500 planes at around $42 Million per plane plus development costs equalling around $200b of total program cost.

When the initial program was finalised in 2002 the unit price had already risen to $50m each.

By 2010, the program's total costs had blown out to around $400 billion, a 100% increase in 8 years.

By 2020, that $400b has blown out to over $500b for acquisition costs alone (now for 20% less planes) and to more than US$1.3 Trillion in total lifecycle costs. The average 3 version cost is now around $100m each.

So much for the *Affordable* fighter plane.

The US Department of Defence, like its counterparts in Australia, continually obfuscates the costs of this program.

Look on any Defence or Defence related website and you will not find details of the original 2001 target estimates anywhere. Try looking for a design variance cost model which shows year by year total program estimated costs vs number of planes to be produced against the original baseline. You will not easily find it.

Look for original estimates of sustainment cost targets from 2001 and you will not easily find them. Look for a sustainment cost comparison that shows JSF is up to 50% higher than comparable planes.

Try and find a robust number for total life-cycle sustainment costs in 2020, a full 14 years after the first flight, and 9 years after production started, and you will not find a definitive concrete number. Plenty of estimates but certainly no relativity back to the original plan. Perhaps that is the result of the blind pursuit of 'bewitching' technology.

So much for the *Affordable* fighter plane.

Try and find the 2001 first estimates of a target cost per hour to fly this plane and you will not find it. Yes, you will find today's estimates, but you will not find any place that Defence or Lockheed measures today's cost per hour relativity with the 2001 original estimate when the US Government first signed on to the program.

Nor will you find a comparison cost per hour to fly that is up to 60% higher than the very planes it is replacing.

So much for the *Affordable* fighter plane.

Try finding the original 2001 target for expected flight availability of the plane in the hands of the Warfighter. Try finding a chart of today's flight availability in 2020 vs the original expected target in 2001.

Australia and 11 other countries signed up to this program without knowing even a firm estimate of the actual purchase price. In 2020, Australia, and the Taxpayer, still don't have any idea what it will cost to sustain these aircraft for the next 30 years.

Price obfuscation has even been reported as extending to quoting a plane's price without an engine. But that is probably more down to our politicians than Defence obfuscation.

The point of all this is to point out why Weapons acquisition programs fail so often.

It does not seem to be anyone's job to make the program deliver on the original promises and targets.

It does not seem to be anyone's job to deliver affordability and producibility in programs.

If cost and time both blowout - then that is just how it is in Defence (this is as good as it gets).

To ask the Dr Phil question: How will that work out for you Australia?

JSF is the acquisition program that just keeps on giving and giving with no end in sight.

I will return to JSF in Chapter Six: The Wall Street view of Defence.

Does this Weapons Acquisition program meet the World's Best Practice standard or benchmark?

No. Only with a very low or very weak standard could JSF Acquisition be regarded as successful.

JSF Result: Acquisition FAIL.

Reader's Analysis:

As we can see the Defence Alternate Way and the Systems Alternate Way describe completely different program outcomes. One success and one failure.

We can now see that the facts and the conclusions of success or failure can be seen vastly differently.

Is the methodology being used to design and develop the JSF Weapon Acquisition 'wholly fit for purpose'?

Reader's Choice:

Now you must choose for yourself which of these alternate realities is where the best truth lies, and which of the alternate realities best describes program success or failure.

Which alternate reality best described a 'wholly (fit) or (unfit) for purpose' Weapons Acquisition program?

Alternate Defence Reality A? or Alternate Systems Reality B?

An Addendum:

What follows is not directly JSF related, but it is an interesting observation about how Weapons programs are decided that in Australia perhaps relates more to the next case study: the 'French folly'.

This reference is based upon an article by Tyler Rogoway, April 5, 2019 in the war zone - "The Only Man Who Flew both The YF22 and The YF23 on Why the YF23 Lost".

It is a story about Paul Metz the test pilot who flew both prototypes (the Y designation) of what later became the Lockheed F-22 program. *Even after flying a pre-production F-22, a far more mature machine than the YF-23 ever was, Metz makes it quite clear that Northrop's offering was on par with Lockheed's, if not superior.*

Metz also noted that both aircraft met the requirements and Lockheed was chosen because the Air Force had greater confidence, they could better manage the program (my emphasis given F-22 / JSF basket-case outcomes).

Metz makes another incredibly valuable point about how Lockheed knew how to present and market their airframe far better than Northrop did. Northrop's team was made up of brilliant engineers - Metz says they were beyond compare - but they thought and spoke almost exclusively in engineering terms. Meanwhile, Lockheed infused far more marketing, salesmanship, and pizazz - 'lasting impressions' as Metz eloquently puts it - into their YF-22 flight demonstration program. They fundamentally understood how to sell their aircraft and how 'showmanship' heavily impacts the acquisition decision-making process.

Northrop didn't, and that fact may have proven fatal for the YF-23.[4]

During my time in Defence people would repeatedly tell me that the Primes were much smarter than Defence. Meaning that they better understand how to 'sell' than Defence knows how to 'buy', and their lawyers were better at framing and wording contracts than Defence was at understanding them.

I will return to this 'marketing' theme's role in Weapons Acquisition programs later in this book.

E. Future Submarine (FS) (aka the 'French Folly')

Folly: noun (stupidity): A lack of good sense, the condition of being foolish, or a foolish action or belief - a way of thinking or behaving that is foolish and likely to have bad results.

Note: my use of the word 'folly' applies only to the choice of builder, and especially their design and development methodology. I make no comment about the technical capability of the proposed French submarine.

For a brief overview of the FS go to Wikipedia: *Attack*-class submarine.[5]

In 2020, this program's costs vary wildly from $50b, to $80b, to $90b, to $145B depending on who you ask, and in what circumstances. This is a truly extraordinary outcome when the Japanese Soryu class would have cost around $15b and the German Boats would have cost around $20b to acquire and build in Australia.

The FS Program to replace the *Collins* class began in 2007 with the commencement of the Defence acquisition project known as SEA 1000. Concept work was originally planned to start in 2009, with a winning design to be identified by 2013 and design work to be completed by 2016, enabling the construction of the first submarine to be completed before 2025. That is, the first new submarine would be available to replace the retirement of the oldest Collins Submarine.

But, by the end of 2014 operational capabilities had still not been defined. In 2015 the Government announced a 'competitive evaluation process' between competing Japanese, French, and German designs.

Footnote: It is instructive to understand why the UK pulled out of the Horizon Frigate program with France in 1999. Any reading of the UK's reason for withdrawing will fill the reader with a strong sense of déjà vu that it could be almost word-for-word a description of what has happened between France and Australia on this program.

FS Alternate Reality A: The Defence Way

On 26 April 2016 the French Shortfin Barracuda, a conventionally powered submarine design based on experience with the Barracuda-class nuclear attack submarine built by French firm DCNS, (now Naval Group) was announced as the winner of the competitive evaluation process (CEP).

The CEP enabled a very thorough analysis of the French, German and Japanese submarine options. A subsequent ANAO report was complimentary about the CEP and Defence's management of it.

Design of what is now known as the *Attack* class submarine is proceeding well and public reports of the impact of relatively minor schedule slippage have wrongly suggested that the program is in trouble. As was expected, settling into a harmonious working arrangement with France has taken a little time but the relationship is very sound and both parties are fully committed to a successful outcome delivered on time and budget. Minor milestone slippages so early in the design process for such a long program can readily be recovered and will not impact the overall schedule. Delivery of the first submarine remains on track for the early 2030s, to be followed by about two years of the normal tests, trials and evaluation before the first submarine is ready for operations. Subsequent submarines will be delivered at about two-year intervals.

Work on new shipyard facilities at Osborne, in Adelaide, are progressing well in close consultation with Naval Group Australia and are on track.

Recent publicity around acquisition costs blowing out are misinformed. Much of the apparent cost escalation is a result of normal exchange rate variation.

The Future Submarine Program is on track to deliver a regionally superior sovereign capability for the security of Australia for decades to come. Thousands of jobs for Australians in many sectors of the economy will be created in the process.

FS Prediction: probability of Acquisition SUCCESS.

FS Alternate Reality B: The Systems Way

On 26 April 2016 the French Shortfin Barracuda, a conventionally powered variant of the Barracuda-class nuclear submarine by French firm DCNS, (now Naval Group) was announced as the winner.

Since then costs have only gone one way, sharply up.

The original $50b was explained as total program costs. By 2020, however, $50b is the cost to design and build only. Hence the $80b number has arisen as it now represents the total program costs ($90b due to exchange rates).

Time has only gone one way: out. That is even before initial design work is 30% completed.

By 2020, the timeline has already slipped to the mid to late 2030's for the first boat in the water. By the time it is fully commissioned and "battle-ready" in the hands of the Warfighter it will be almost 2040 at the earliest, assuming no more delays and schedule slippage occurs. But history suggests that is indeed a faint hope.

Therefore, it will be another 20 years before we have one new submarine battle-ready and 26 years before we have a small fleet of 3 new submarines fully battle-ready in the hands of our Warfighters. That is, the first boat will be ready a full 33 years after the need was first recognised. While a battle-ready fleet of three will take 39 years.

How did Australia get to this point?

How on earth could a program go from contract and approval in 2018 to potential basket-case in less than 2 years?

What was our Government and Defence thinking?

Did French 'marketing bullshit' beat Australian brains?

Clearly affordability was never considered as a factor.

Clearly producibility was never considered a factor.

Defence will vociferously dispute all these figures. But that is because Defence has never issued a solid schedule of costs for each separate component of the program. There has only ever been a very 'muddy' number that can be fudged to suit any situation. Obfuscation by Defence of cost and schedule details is entirely the norm in Weapons Acquisition programs.

To ask the Dr Phil question: How do you think this Acquisition program will work out for you Australia?

- 33 years from first concept to 'battle-ready' as a weapon of war.
- $30b more than the next lowest competitive tender (before any current or future blowouts).
- $35b more than the lowest tender.
- We are currently less than 6% of the way through the program.

Does this Weapons Acquisition program currently meet the World's Best Practice standard or benchmark?

No. Only with a very low or very weak standard could FS be predicted to be successful.

FS Prediction: Extreme Probability of Acquisition FAIL.

Reader's Analysis:

As we can see the Defence Alternate Way - and the Systems Alternate Way - describe completely different probability program outcomes. One predicts success and one predicts failure.

We can now see that the facts and the conclusions of success or failure can be seen vastly differently depending upon the perspective under which they are written and analysed.

Is the methodology being used to design and develop this FS Weapon Acquisition 'wholly fit for purpose'?

Reader's Choice:

Now you must choose for yourself which of these alternate realities is where the best truth lies, and which of the alternate realities best describes program success or failure.

Which alternate reality best described a 'wholly (fit) or (unfit) for purpose' Weapons Acquisition program?

Alternate Defence Reality **A?** or Alternate Systems Reality **B?**

F: The Common Denominators of Failure

All of these past problems in acquisition programs share a common dynamic: - moving forward with Weapons Acquisition programs **before** all of the necessary and correct data and knowledge needed to make the right decision is available at every single program decision point, and at the precise time it is required.

This always results in a highly **reactive** acquisition methodology which more resembles a 'whack-a-mole' random thinking problem solving process, which will **always** deliver cost and time blowouts.

The common denominator across all five case studies is the critical lack of three critical ingredients.

1. Manufacturability

Manufacturability describes how easy or how hard something is designed to be manufactured. To determine how manufacturable any design is, we often use a technique known as the manufacturing complexity penalty score.

The more complex or difficult something is to develop, make and assemble, the higher the penalty score. The aim in every good design methodology is to reduce this penalty score to the absolute minimum, or zero if possible.

In a Toyota product, they can typically only reduce this penalty score by around 10-20% because they have already made every previous design easily manufacturable. In general industry, a firm can be taught to easily reduce this penalty score in their design by around 45-70%. However, in Defence, 80-90%, even 95% reduction in the penalty score is often achievable because Defence does not ever require contractors to design for manufacturability.

Yet as can be seen with these case studies one of the key failure modes is the extreme difficulty of the local manufacturability of the ships, submarines or planes.

Manufacturability is a key missing ingredient in all these Defence Weapons programs.

2. Producibility

While manufacturability looks at how manufacturable each component is, producibility looks at how producible the ship, submarine or plane is as an entire battle-ready weapons system.

Defence does not use producibility modelling in Weapons Acquisition programs, as these case studies show.

Producibility modelling is sometimes used on defence programs in the US. When it is, the result is often a long way below 50%, a truly bad result out of a possible score of 100%. A good producibility modelling result is anything above 90%. Highly competent firms often achieve >95% results.

Producibility modelling is the second key missing ingredient in all these Defence Weapons programs.

3. Affordability

Without highly controlled manufacturability and producibility, affordability is not possible. But it also works the other way around. Good affordability modelling mandates good manufacturability and good producibility outcomes.

Good affordability targets also mandate high levels of Innovation to deliver manufacturability and producibility.

Defence does not use highly robust affordability modelling in Weapons Acquisition programs.

Affordability modelling is the third key missing ingredient in all these Defence Weapons programs.

G. The Final Word.

From **Sir Bernard Gray**, Chief Executive Officer of Defence Matériel, MOD UK.

Excerpt from - Public Accounts Committee - Minutes of Evidence - Monday, 20 May 2013

Bernard Gray: There is the setting of requirements and then there is delivery of the equipment itself. While I appreciate that the defence industry will quite often say that it wishes to be left alone, thank you very much, my experience is that that is not, on the whole, a good idea. It is fair to say that on most occasions when I have pushed on specific issues, they are not as well covered off as they should be. If I just let a contract and walked away and invited defence contractor A to get on with it and

"Do just please drop by and deliver the equipment at the end of it and I'll write you a cheque", *I am unlikely ever to get that equipment.*

Q102 **Ian Swales:** Why?

Bernard Gray: Because their control of programmes is not all it might be.

Q103 **Ian Swales:** So, we have to get involved in controlling the programmes of our suppliers? Is that it?

Bernard Gray: If I can take you back to the most salient example of this, in the Astute programme we did what you suggested. (leave BAE alone) *It was a disaster.*

From 1996 to 2003 we let them get on with it. We had a contract and that is what we cared about.

In 2003, it almost broke BAE Systems. It cost them hundreds of millions of pounds. We then had to step back in, reformulate the programme and effectively recuperate the whole of our submarine-building activity, *which is something that is only beginning to come right some 10 years after that disaster.*

Note: The head of UK Weapons Acquisition's words are describing the incompetence view of Defence. He is saying that the internal processes of the primes are so unsatisfactory that without strict oversight they cannot deliver the promised weapon capability into the hands of the Warfighter.

Taken to its extreme, this suggests that eventually the Warfighter would run out of equipment to prosecute war.

Also ...an excerpt from - Pentagon's Office of the Director, Operational Test & Evaluation (DOTE) FY13 report...

...Looking at three decades of reliability assessment, the report concludes that *"the reliability of DoD systems has not significantly improved over time"* whether under MIL-STD-785B in the 80s-90s, following commercial best practices in the '00s, or under the current, more prescriptive policy.

"The reasons programs fail to reach reliability goals include inadequate requirements, unrealistic assumptions, lack of a design for reliability effort prior to Milestone B, (cut steel) and failure to employ a comprehensive reliability growth process."

NOTE: DOTE is saying that over 30 years of Acquisitions, things have not changed for the better.

This suggests that DOTE is also describing the Incompetence view of Defence.

No matter what reforms or tougher standards are enforced - the reliability results have not improved.

H: A Metaphor

Let me use the following metaphor to explain the two alternate realities.

The Aim: To get to Sydney from Melbourne the best way (Build the best Warship).

The Defence Alternate Reality: set out from Melbourne - end up in Perth - then to Darwin - then to Brisbane and finally to Sydney. So, it has taken more than twice as long as it needed to - and cost more than twice as much as it should have, but we finally arrived at the planned destination.

Defence is now saying to you: "What are you complaining about? Just look at the fine city." (Warship)

"Cost? Time? They are immaterial - they are just the price you have to pay for such complex decisions". "Getting from Melbourne to Sydney very effectively is really hard, so you must always expect delays and diversions".

The Systems Alternate Reality: How you get from A to B actually matters. Respect for the Taxpayer actually matters. Incompetence is no excuse for delays and diversions. Incompetence is never an excuse for not delivering the best possible outcome for all concerned. 'This is as good as it gets' must never be acceptable.

> *"If a sufficient number of management layers are superimposed on each other, it can be assured that disaster is not left to chance."*
>
> *~ Augustine's Law Number XXVI*

I. A Robust Proposed Solution:

Solving Defence Weapons Acquisition System Core Problems:

The solution to these problems is to only use a highly **predictive** methodology that **always** delivers all of the correct and necessary data and knowledge needed to make the right decision, and which is available at every single program decision point, and at the precise time it is required.

Affordability, Manufacturability, and Producibility must be mandated.

Mandating this solution requires that only a <u>highly predictive</u> methodology must be used in every future acquisition program.

In the following chapters a new methodology and capability will begin to emerge which I will term WBP2020 (World's Best Practice 2020).

In Part 2, WBP2020 is proposed as the best solution to drive weapons acquisition reform.

J: Readers' Choice

You have now seen the case studies, the common denominators, final word, a metaphor, and a solution.

Do you think the Alternate Defence Reality Way or the Alternate Systems Reality Way describes where the best truth lies and which of the alternate realities do you think best describes program success or failure?

What is your overall choice?

Alternate Defence Reality **A?** or Alternate Systems Reality **B?**

Reader's Choice: Revisited...

Before we leave this chapter, let me ask you the reader a far more direct and personal question.

If this were your own personal money being spent, would you choose the current Defence way of weapons acquisitions, or would you prefer to spend your money on a tried and true 'world's best practice' weapons acquisition methodology?

Case Study Summary:

My job in these case studies has been to lay out the best facts as I see them. Defence has its own alternate facts as it sees them. It's now up to you the reader to decide among these alternate realities where the best truth lies.

Do these case studies portray a weapons acquisition system that is 'wholly fit for purpose', as Defence claims, or do they portray a Weapons Acquisition system that is 'wholly unfit for purpose'?

In my view these case studies clearly demonstrate outcomes that are 'wholly unfit for purpose'.

Let me use a truism from the commercial world that also seems to apply in Defence.

'Marketing bullshit' is what you get when all too often what gets chosen is not the strongest or best idea - but an average idea presented in the slickest way.

Defence keeps perpetuating the same 'known' problems and consequences again and again.

Unfortunately, success in Defence today is about spending all the available money and then more - not about better outcomes for the Taxpayer or Warfighter.

So, the next time you hear Defence declare success - you should very bluntly ask - compared to what?

A Final Note: there will be a lot of snorting and puffing by Defence and its many cheerleaders around the financial numbers used in these case studies.

But you must not let them confuse you.

Instead, ask them why they have deliberately chosen to not reveal the official detailed Cost Table which will clearly display the entire history of every cost movement starting from the initial concept right through to the battle-ready completion of every project. (For reference a very simple Cost Table example is shown in the AWD Case Study).

If there is no Cost Table shown and discussed - then you can be reasonably assured that none exists.

And the puffery and snorting are simply a means to obfuscate that truth from you.

---

1. *https://en.wikipedia.org/wiki/Collins-class_submarine*
2. *https://en.wikipedia.org/wiki/Hobart-class_destroyer*
3. *https://en.wikipedia.org/wiki/Lockheed_Martin_F-35_Lightning_II*
4. *https://www.thedrive.com/the-war-zone/27309/the-only-man-who-flew-both-the-f-22-and-the-yf-23-on-why-the-yf-23-lost.*
5. *https://en.wikipedia.org/wiki/Attack-class_submarine*

4

A FRAMEWORK OF REFERENCE

(TO EXPLAIN WHY THIS SHIT HAPPENS)

Frame of Reference: A set of distinct criteria, stated values or a standard, in relation to which measurements or judgements can be made about an event: i.e. a baseline to make a comparison in relation to events.

For the reader's benefit it is necessary to first establish a base framework of reference, or standard, against which the following chapters (and the preceding case studies) can be properly measured and judged.

Every Product Development program MUST address what are known as Trade-offs.

So, if Trade-offs **MUST** happen, then we must address two vital questions.

Q1. What are Trade-offs?

Q2. When in the development process MUST these Trade-offs be resolved?

1. What are trade-offs?

These Trade-offs include (but are not limited to) quality, cost, time, manufacturability, technology integration, program affordability, and sustainment (the ability to service, maintain and upgrade over the equipment's life).

For example, all quality problems or issues will have cost, time, manufacturability, and sustainment impacts.

A Simple Example: The Make vs Service Trade-off.

a) Easy to make + easy to service
b) Hard to make + easy to service
c) Easy to make + hard to service
d) Hard to make + hard to service

The ideal is of course Trade-off outcome a). But Defence most often delivers outcomes of d, or sometimes c, Never b, Never a.

This explains why most Defence programs have very high sustainment costs vs costs in other countries.

2. When in the development process are these trade-offs resolved?

There are only 5 places in any Weapons Program where trade-offs can be resolved.

1. During concept design phase: i.e. before any design drawings on paper, or in Computer-Aided-Design (CAD).
2. During CAD or paper design phase: i.e. before we begin to 'cut steel'.
3. During the EMD phase: i.e. during engineering & manufacturing development and before production starts.
4. During production phase: any unresolved Trade-offs left over will be found as production problems.
5. During military deployment: any unresolved Trade-offs left over will be found by the Warfighter.

Obviously, the earlier in the process these trade-offs are resolved the better for the whole program.

A. **Predictive:** e.g. **WBP2020:** around 90% of trade-offs are already resolved prior to drawing in CAD. ~ 95% have been resolved by the time we 'cut steel'. ~99% have been resolved by the time we go into production. And we can expect 0% are passed into the hands of the Warfighter.
B. **Proactive:** e.g. **Toyota:** around 70% are resolved prior to drawing in CAD. ~ 90% have been resolved by the time we 'cut steel'. ~97%

have been resolved before we go into production. ~2% are still to be resolved by production. ~1% are passed on to the customer as warranty or service problems.

C. **Reactive:** e.g. **AWD:** only around 20% are fully resolved prior to drawing in CAD. ~ 50% have been resolved by the time we 'cut steel'. ~80% have been resolved before we go into production. ~10% are still to be resolved during production. ~10% are passed into the hands of the Warfighter.

D. **Basket-case:** e.g. **Collins:** only around 15% are fully resolved prior to drawing in CAD. ~ 40% have been resolved by the time we 'cut steel'. ~60% have been resolved before we go into production. ~20% are still to be resolved during production. ~20% are passed into the hands of the Warfighter as operational, or sustainment (service, maintain, upgrade) difficulties.

The ideal is of course, A.

Affordability and producibility can only be delivered by A, partly by B, Never C, and Never D.

But Defence most often delivers D or sometimes C. Never B and Never A.

This explains why most Defence programs have very high cost and time blowouts vs other countries.

Right First Time.

Right first time is a predictive or proactive way of thinking.

In Toyota it is known as 'front-loading'. The thinking works like this...

Resolving Trade-offs, and the problems they cause, before we draw the design on paper or in CAD costs nothing.

But, after CAD the rework/reissue of drawings costs a little. Once we cut steel, any rework/redesign costs now escalate quickly, and cost blowouts begin. During EMD (engineering & manufacturing development) costs of making changes are now very high. Making design and engineering changes once in production becomes cost prohibitive. Making changes in the hands of the Customer/Warfighter is by far the highest cost of all.

So, **right first time** is a way of thinking that says the lowest cost design is one in which all major trade-offs and problems are resolved as early as possible in the design and development process. The earlier the better.

WBP2020 is a right first-time model for lowering the cost of future Weapons Acquisition programs.

This is why WBP2020 can reduce the cost of weapons programs by as much as 50%.

Right Second Time... or third time, or fourth time

This is a faulty and dysfunctional way of thinking based upon the idea that...

> *"There is no time to do things right the first time, but all the time we need will be made available to do it right the second time. (or third, or fourth time) So multiple design changes made after we cut steel are fully acceptable."*

Design changes made during the EMD phase are completely acceptable and design and engineering changes in production or in the hands of the Warfighter which dramatically blowout costs and timelines are fully acceptable.

This is typical of an entirely reactive framework of thinking that permeates the way defence thinks today.

The prior case studies all describe a 'right second time' mindset that is completely acceptable to defence.

This is why cost blowouts (and time blowouts) of 50% to100% are normal in Weapons Acquisition programs.

My Experience in Defence (anecdotally)

When I first started working in Defence I was told: "Oh, Toyota? Well, in Defence you will quickly learn, there is the right way, and then there is the Defence way."

And never a truer word was spoken. The Defence way of thinking dominates everything to the exclusion of every other way of thinking. It is so pervasive that it is a key reason behind why Defence is so risk averse today.

But it's also hard to blame the people inside Defence. They have been through almost constant reviews and reforms over the past 30 years. I lost count at 19 reviews in the past 25 years and gave up.

Why so many reviews with no visible improvement?

Defence Weapons Acquisition programs have been a serious problem since the 1990's.

The two Adelaide class Frigates (1985-89) took only 4 years from keel laying to commissioning and the Anzac Frigates (1993-03), which took 3-4 years from keel laying to commissioning, were the last low problem programs to be built in Australia at Williamstown shipyard.

With Collins things started to go awry.

The first boat took 6.5 years from keel laying to commissioning (18 months late) while the last took 8 years from keel laying to commissioning (41 months late) and program costs went from A$3.9b to A$6.2b.

Three Anecdotes:

I spent 10 years trying to get anyone from Defence interested in the WBP2020 methodology (or its predecessor versions) as described later in this book. I will use three anecdotes to describe the responses I received.

I was approached at a Pacific Naval conference by someone I knew well from Defence Acquisition who said:

"What are you doing here?"

Me: "Talking to people"

"No, I mean - What are you **doing** here?"

"Can I give you some free advice?"

"Why don't you fuck-off out of Defence and save yourself a lot of heart-ache."

"You've been talking to Defence for the last 5 years and getting absolutely nowhere."

"It's always been a one-way conversation. You always call us, but we never call you back."

"If we (Defence) had any interest in what you can do for us, we would be all over you like a rash."

"But we don't – so why don't you do yourself a big favour – fuck-off out of Defence and stop trying to fix something that Defence has no interest in fixing."

I eventually took that as great advice and gave up trying to work with Defence and the Defence Primes.

But I retained membership of the supplier network so I could continue to talk to Defence SME's.

Two years later, I had failed to get any of those Defence SME's to act.

Then I asked **five** SME CEO's who I knew well enough for an honest conversion about why Defence SME's were not interested despite their own Engineers expressing interest in understanding the WBP2020 methodology.

The first CEO fobbed me off with general platitudes.

When the second CEO began that same response, I stopped the conversation and told him I needed a brutally honest response from him – because I needed to understand whether I was wasting my time by continuing to persist with talking to Defence SME's about WBP2020.

He paused for a few moments – then said:

"OK – but none of this can ever come back to me."

"My people have talked about the long-term benefits and value of what you are describing but, as CEO, why would I put my company and my people through the heartache and struggle of learning such a different methodology and its very different way of thinking, when what I am doing right now is perfectly acceptable to Defence."

And there you have it. The enemy of change. "What I am doing right now is perfectly acceptable to Defence."

Cost blowouts, time delays, program problems are 'acceptable'. 'This is as good as it gets' strikes again.

The low bar standard or benchmark in Defence wins again, and again, and again.

BTW - when I put this position to the other 4 CEO's they all concurred with the sentiment expressed above.

As a final anecdote, around two years ago I spoke to a former Defence Prime CEO who had retired a number of years previously. I asked him why Defence Primes like the firm he had commanded had shown no interest in better development methodologies like WPB2020 (or its predecessor versions).

He replied that they saw activities like WBP2020 as a 'Profit Reduction' threat. The conversation got a bit vague at this point - and he insisted upon only speaking 'generally'.

Defence Primes seem to think that using a new method like WBP2020 which could reduce the total actual cost of a program by say 30% - would in-turn reduce the firms revenue stream by 30% - which in-turn would reduce profit dollar ($) amounts by around 30%.

Note: actually, this is a correct interpretation of adaption of WBP2020.

Although it was unstated, this also implied that the CEO's bonus which is, of course, based on increasing revenue and profits would be reduced, or even eliminated if the CEO of a Defence Prime did implement WBP2020.

This was not the first time I had heard the term 'Profit Reduction' activity. In fact, a non-Defence CEO once said straight to my face that he would not talk to me because he said WBP2020 was a 'Profit destruction machine'.

And so, the Wall Street view of Defence strikes again.

Perverse incentives always deliver perverse results.

Too bad for the Taxpayer.

Too bad for the Warfighter

But, of course, the opposite is also true.

By adopting WBP2020, any firm could secure a fast-growing future in their marketplace by delivering far better products and equipment that are far more affordable, with far better manufacturability and producibility, and with far less quality, sustainment and operational problems for the Warfighter.

One CEO I know who has taken this path said, "I know my competitors will NEVER adopt WBP2020 – so, if I do then I can eventually take all of their business away from them."

Note: this too is a correct interpretation of adaption of WBP2020.

This way of thinking comes from Toyota. Toyota requires its suppliers to be profitable so they don't need to cut-corners in ways that may impact the quality or performance of the parts they supply to Toyota.

For example, in 2006 Toyota wanted to put a luxury feature from the Lexus on the Camry. But, of course, the Lexus part itself is too expensive to put on a Camry. Toyota then pulled together a team of firms and academics who worked out how to design, develop and manufacture the required feature for an affordable price that Toyota was prepared to pay, with an acceptable profit margin for the supplier.

5

THE INCOMPETENCE VIEW

Incompetence (*noun*): inability to do something successfully, or as it should be done; ineptitude. The lack of the ability, skill or knowledge that is needed to do a job or perform an action correctly or to a satisfactory standard.

Incompetence is a very hard-nosed word and it should never be used lightly.

But what other appropriate word can anyone use to describe the preceding case studies apart from incompetence?

How can anyone possibly justify that level of dysfunction, cost blowouts, time delays and capability missteps without labelling it as 'incompetence'?

I had a career spanning 23 years working across the Toyota Group of companies. Within Toyota there is an almost maniacal focus upon doing the absolute best possible job for the customer. It is a focus which abhors incompetence, ineptitude, or waste in any form. For people like me, high-performance capability like WBP2020 is absolutely normal. So normal, that we find it hard to understand how anyone could continue to use vastly inferior methods. Yet, when we go out into general industry, we find that methods like WBP2020 are extremely rare.

And in Defence and the Defence supply chain, rarer still.

Why is the organisational incompetence model so persistent in Defence?

Simply because: for many, it still works extremely well.

Organisational Incompetence was and is always, a failure of leadership.

30 years ago, I was taught by a very Senior Executive from Toyota how to 'see' incompetence in leadership.

*A **warning**: what follows is metaphor that is very politically incorrect.*

But it is simple and works brilliantly.

During a factory visit, and before any explanation from the management, he always insisted upon a factory walk.

This Japanese Executive explained that 'shit' only ever flows downhill. i.e. from the top ('shit' being a metaphor for problems, broken systems, dysfunction and stress on the people doing the actual job).

As he walked around, he would imagine how deep in the 'shit' the employees were standing.

Was it up to their ankles, up to their waist, or up to their necks?

With this mental image, or snapshot, he was now ready to listen to the management.

If the 'shit' was only at their ankles he expected to hear a management which downplayed itself. This was a management that knew what it was doing and he could expect that what he would hear would be in line with what he saw.

But if the 'shit' was up to their necks, he expected to hear a management which talked itself up. This was a management that had no clue what it was doing. Had no clue how to run a successful business. And the management presentation was nothing like what he had already seen with his own eyes.

I have used this technique myself for the past 30 years and can attest to how accurate these 'snapshots' are.

This metaphor comes from what in Toyota was called 'Management by Stress'. Management by Stress refers to any situation where they 'stress the human to make up for the failed inadequacies of the system'. The correct approach in Toyota of course, is to stress the system so we don't stress the human.

As the CEO of a Toyota company my job is really simple. My job is to reduce the level of 'shit' my employees are standing in, down to the level of their ankles (there is never no stress on the people).

10 years ago, after a talk in the USA and during Q&A, a young man stood up and said. "You just described my boss to a T, what should I do?"

"You should leave," I advised. "The stress on the people from such a boss who has no clue what they are doing will only ever increase until you can no longer tolerate it. So, leave now."

A year later I received a letter from the young man informing me things had changed and he had stayed. It transpired that when he returned to work a board member had asked him what key insights he had learned at the conference. He recounted my talk and my response to his question. This in turn created a conversation within the Board. The culmination of which was that the CEO had been sacked as part of a management clean-out.

It seems that God really does move in mysterious ways.

But it drives home the point that incompetence is always a leadership problem.

So it is with Defence. Defence Acquisition is a 'Management by Stress' system. It relies heavily upon the heroic intervention of the individual people buried deep inside the system, to rise to the occasion so as to minimise as much as possible the damage from the built-in dysfunctional inadequacies of the system.

It's never the fault of those working deep in the bowels of the system. It is only ever the fault of leadership.

Incompetence is always, always, a failure of leadership.

And then leadership incompetence creates organisational incompetence.

Note: for those readers further interested this subject, I will deal specifically with it in Book #3. Book #3 deals directly with the question "Why does nothing ever change?"

This aversion to change is not just true in weapons acquisition (this book), or in our national Innovation systems (the subject of Book #2), but it also explains why the systemic organisational failures from incompetent leadership are so deeply pervasive across all aspects of our Political, Bureaucratic, and Industry leadership landscapes.

6

THE WALL STREET VIEW

WALL STREET: The single-minded focus in which the *share price* of any firm is the *only* thing that matters.

Others might suggest it means Gordon Gekko's "greed is good" credo which typifies the short-term view.

I was first introduced to the Wall Street view of Defence following an address to an MBA class a few years ago.

In my presentation I used the AWD and JSF case studies (plus non-Defence) to show why incompetence causes cost blowouts to happen in all forms of development programs and products (not just in Defence).

That is, I was explaining why affordability and producibility can never happen in these programs.

After my presentation was finished, I was approached by two MBA students who said, "You know you're wrong about JSF."

Me: "Oh?"

"We attended an MBA school in the USA where JSF was one of the case studies."

"Using the same data that you presented: 2002 Initial order intent of 3500 planes at around $42 Million per plane plus development costs equals approximately $200b total revenue and $25b profit."

"But by (2020) that $200b has blown out to over $500b for acquisition costs alone (for 20% less planes) and to more than US$1.3 Trillion in total lifecycle costs - at a total profit now of around US$200b"

"Now isn't that a *fucking* brilliant business model!"

Me: (speechless)

It took me a very long time to come around to acknowledge the validity of this Wall Street view of Defence simply because the implications behind that blatant emphatic statement were initially too shocking for me to accept.

But following their advice, I began to follow and map the share performance of a number of key Defence Primes. And sure enough, I saw that actual program performance had no correlation to share price movements.

If anything, it seemed share prices moved inversely to the actual program performance delivered. So, the worse the actual program seemed to perform, the higher the share price went.

Finally, I have had to concede that their Wall Street model of Defence was indeed valid. Shocking as that may be!

It seems for some that the single-minded focus where **share price** of any firm really is the **only** thing that matters.

Too bad for the Taxpayer.

Too bad for the Warfighter.

Share Price rules!

A Practical Example: The AWD.

Someone made an absolute shedload of money out of the AWD program. The Government planned to spend only $4.8b of Taxpayer money. But ultimately, they will spend more than $10b. Where did the extra $5b go?

The $5b went to every firm who contributed to the incompetent design, development and build of these ships.

But none of that extra $5b went to any of those few good suppliers who used robust design and development methods that delivered good, high quality components on time and on cost.

In a sense, all of the extra $5b went to the 'bad' suppliers who made continual mistakes. The 'worse' the supplier delivered in terms of performance, the more they shared in the windfall $5b of taxpayer money as profits.

There are NO rewards for competent cost reduction or better methods that will deliver better equipment.

A question to the reader. If you are the CEO of a 'good' supplier what should you do? Continue to 'miss out' on a share of $5b. Or do you decide to become just another firm more focussed on share price and profit self-interest?

These sorts of perverse incentives are one reason why nothing ever improves in weapons acquisition programs.

Too bad for the Taxpayer.

Too bad for the Warfighter.

But the Wall Street view does not end there...

Both the incompetence model and the Wall Street acquisition model above create far bigger whole-of-lifecycle affordability problems known as the Sustainment Trap and the Maintenance Repair & Overhaul Trap.

The Sustainment Trap: (service, maintain, upgrade)

A Sustainment trap **always** arises in every reactive or basket-case Defence Weapons program.

This is because there are so many problems during design and especially development that no time has been spent to make the equipment's sustainment (serviceability) affordable. Often in these programs sustainment is not even thought about until just before the equipment is delivered to the Warfighter for commissioning.

By then, the cost of sustainment is locked in and cannot be reduced without going back and changing the design. But, of course, we are now far too late in the development to allow such significant design changes just to reduce sustainment costs. Catch-22. Hence the complete un-affordability of most sustainment programs.

A Real Example: About 9 months out from the handing over to Navy of the first amphibious assault ship (HMAS Canberra, known as an LHD or Landing Helicopter Dock which is the biggest ship ever operated by the Australian Navy, and which was also reintroducing multiple embarked air craft operations from a single ship, as well as a modern heavy-electrical propulsion system the like of which Navy had never operated before) there was no agreed sustainment concept for those two ships. It's not that they hadn't thought about it - they had - but they had no idea how they were going to approach sustainment. Essentially the same situation was faced with the AWDs.

Compare this with WBP2020 where the initial sustainment readiness level (SRL) and sustainment affordability level is known at the end of the concept design phase and before anything is drawn in CAD. The SRL is then verified and maintained for the whole of the development process, right thru until it is the hands of the Warfighter.

Now we have to consider this from the Wall Street view. Is it in the best (profit) interests of the prime contractor to use SRL in WBP2020 or to continue to use the incompetence/Wall Street model? We have already seen back in chapter 4 that WBP2020 is seen as a 'profit reduction' threat to the Wall Street model of defence.

Typically, it is in the best financial interest of the prime contractor NOT to use a predictive system like WBP2020.

Typically, they will make far more profit from reactive or basket-case development programs, where they are now paid for 30-50 years to fix problems that otherwise could have been eliminated with a predictive methodology.

The MRO Trap: (Maintenance, Repair & Overhaul)

MRO is another question about what is in the best financial interests of the Defence contractor.

With modern Defence equipment becoming so complex because of electronics and software, the prime contractor often demands exclusive rights to repair that equipment for the whole lifecycle of that equipment. That is, only the manufacturer now has the depth of knowledge to competently repair any faults or warranty issues.

Often these contractors will even claim monopoly IP rights to new technologies paid for by the Taxpayer.

Which may, or may not, be ok when the equipment is at its base or a proving ground, but what happens in a war situation? In a battle-zone there is no-one who has the knowledge to attempt to repair the equipment in-situ. Even any military mechanics with the battle group will have had no experience fixing the equipment because it must be fixed by the manufacturer, but the manufacturer will not send its experts into a battle-zone. So, Catch-22!

So how does it get fixed in a battle-zone? Simply, it does not.

Either of two things will now happen.

One, they abandon the military mission altogether (because they can't leave the equipment out in the badlands) and tow it back to base, or two, if the mission is critical enough, they will blow up the multi-million dollar piece of equipment so it can't fall into the hands of the enemy, and move on with the militarily important mission.

This MRO trap has a commercial world equivalent with farm tractors. Tractors are now so complex that most manufacturers demand exclusive rights for all warranty and repairs for the operating life of the tractor. These monopoly rights mean there is no competition for repairs. No competition always leads to higher costs to repair.

Typically, it is in the best financial interests of the contractor to demand exclusive repair and overhaul rights.

But what is in the best interests of the Taxpayer or the Warfighter to help them to successfully prosecute a war?

The 'Profit Reduction' Trap:

When I later started to ask questions along the Wall Street view no-one would directly admit that it was a valid view.

But there always seemed to be a 'nudge-nudge-wink' acceptance that it was a real view - even inside Defence.

This 'Profit Reduction' trap must not be ignored as it implies that anything that is seen to reduce a firm's revenue and profits from basket-case programs will be resisted as it is not in the firm's best interests to use better methods.

Therefore, it is never in the firm's best interest to do activities like predictive sustainment.

This implies that knowing cost per hour to fly a new plane (or cost per day per ship, or cost per kilometre to operate land vehicles) at the start of the program is against the firm's best interests. That is, it is far better that this is not known until very late in the program when it is far too late to do anything about it.

This implies that affordability & producibility outcomes known up front is not in the best profit interests of any new development, remediation program, sustainment needs, or any MRO needs.

The Wall Street view is all pervasive.

The 'Sacred Cows' remain sacred.

Is it Wall Street's fault?

Now this all sounds like all Defence Prime contractors are villains who are in it solely to make obscene money.

Perhaps they are, but we do have to ask the question of whether this is all their fault, or is it the fault of the Defence bureaucracy who continue to allow reactive and basket-case programs to happen time after time?

Perhaps the real question is not about why the Primes don't use much better methodologies like WBP2020. After all, without very strong oversight, it is in their own financial best interests to always maximise profits.

Perhaps the real question is why the Primes are much smarter than Defence - certainly smarter at Marketing to Defence - and why they use smarter and better lawyers and negotiators.

Perhaps the real question is why our Politicians and Bureaucrats have no interest in introducing greater affordability, producibility and sustainment methodologies with robust oversight methods into every defence weapons program that would completely eliminate all reactive and basket-case weapons development programs.

Perhaps the real question is why no-one is focussed on affordability and producibility of weapons programs.

Perhaps the real question is why no-one seems to be focussed upon the critical need of the Warfighter to always be ready to successfully prosecute war with the best possible equipment available to them all-day, every-day.

Wall Street View summary:

*The single-minded focus where **share price** of any firm is the only thing that matters.*

This chapter has discussed a view whereby cost blowouts and long-time delays are not bad, but in fact are good for the profitability and share price of the individual Defence firms involved.

We cannot ignore the impact that the Wall Street view has upon Defence Weapons Acquisition programs.

It is a very valid business model for every Defence Supplier.

Cost blowouts are excellent for the whopping share prices of Defence supply chain firms

But of course, what's good for the Individual Defence Suppliers is not in the best interests of the Taxpayer.

Because it can have severe impact upon the Cost and Time requirements in Weapons Acquisition programs.

For me, this is simply the incompetence view observed from another angle.

The Incompetence view simply allows the Wall Street view to happen.

Either way there are some obscene amounts of Taxpayer money wasted in Defence Weapons Acquisitions.

This is a question you as the reader will need to consider for yourself as the rest of Part One unfolds.

Finally, no-one should underestimate how difficult it will be to change the status quo in the Wall Street view of Defence Weapons Acquisition and the absolute determination to retain the 'sacred cows' that exist today. Resistance to change will be fierce and persistent despite the apparent ease of implementing the solution as described later in Chapters 15 to 19.

This warning should not be ignored by anyone.

7

WHY THINGS NEVER CHANGE

The purpose of this Chapter is to set the scene for the following chapters so you the reader can get a good feel for why nothing will change in weapons acquisition **until the current model is disrupted.**

A more detailed understanding is explained in Part 2 of this book.

Introducing the Evolution of World Best Practice 2020 (WBP2020):

World Best Practice 2020 (WBP2020) is a convergence of ideas over the past 10 years into a single methodology designed to deliver more affordable products or services with much higher quality and lower costs.

WBP2020 is a hybrid of the Toyota Development System (TDS) and the Munro Development System (MDS).

I was first taught the basics of TDS in the 1980's while working in the largest of the Toyota Group companies.

I was first introduced to MDS in 2001 when Sandy Munro visited Australia to teach MDS to Automotive suppliers. At that time only a small handful of suppliers took up the MDS capability. My firm was one of those.

Being a Toyota company, we were never allowed to use MDS on any Toyota products, but we did use MDS on Ford, GM Holden, and Mitsubishi products. Using this early version of MDS we were able to reduce the cost of products by around 30-50%. We were so successful at reducing both cost and time by using MDS that we were able to avoid off-shoring our products to China as almost everyone else in the automotive supply chain was doing.

While this was going on Munro & Associates was working with the US Office of the Secretary of Defence (OSD) under Defence Secretary Robert Gates on a reform program that culminated in legislated reform in 2009. OSD and a number of other US defence think-tanks had begun working on ideas around how to stop their basket-case weapons development programs.

Initially OSD had asked if Munro's MDS methods could move a major program decision point from two-thirds of the way through the EMD phase (engineering & manufacturing development) back before Milestone B (Cut Steel), i.e. move from a reactive model to a proactive model. For Munro this was relatively easy because, like TDS, MDS was already capable before that cut steel decision point.

The result was legislated by Gates in 2009.

Buoyed by this easy success, OSD began asking if MDS could now be modified to bring that major decision point back before milestone A (draw in CAD). i.e. create a fully predictive model. Over the next few years OSD funded a number of development trials aimed at proof of concept for this new requirement.

Gates' aim was to force Weapons Acquisition Programs to move from reactive to proactive and, finally, predictive.

Eventually Gates' requirement was met. But unfortunately, this coincided with his retirement as Defence Secretary. Subsequently, the program was abandoned and the US returned to a more traditional reactive weapons purchasing model.

But the work was not lost. Munro simply moved this newly developed capability into the commercial world.

When I finally left the Toyota group to set out on my own, I picked up this new Munro capability and began morphing it together with my previous TDS experience with the aim of offering this new combined capability on Australia's new future submarine program know at that time as SEA1000.

I then requested Munro to help me to create a specific SEA 1000 version based on their Gates era work and morphed this together with my TDS knowledge to create WBP2013, the first version of what is now WBP 2020.

After making good initial progress I began to run into that same old wall that everyone else runs into.

Later still, the SEA1000 program was overhauled, my work was sidelined in the restructure, and then dismissed in favour of the 'normal' reactive Defence weapons acquisition way which quickly reassumed total control over that program.

For the reader, this was where the, *"What are you **doing** here?"* anecdote found in Chapter 4 came from.

So, just like Munro I abandoned my attempts to work with Defence and moved my focus to the Commercial world where I continued to develop the final version into WBP2020 as it exists today.

Note: I don't expect the reader to comprehend the intricacy of this model. Merely to understand that it is a highly predictive methodology designed specifically to substantially reduce the cost of products, programs or services, while simultaneously shortening the time taken to develop and introduce those products, programs or services.

At its core is affordability and producibility.

More will be explained in Part 2 of this book.

Introducing Cost Reduction: Probably the single most misunderstood concept of all activities.

First a narrative:

Look around you and find a small, simple product around the office or the house, like a desk calculator or a small weather station that costs around $20. Pick it up. Look at it. Study it.

Now ask yourself this question. How could I reduce the cost of this by **at least** 50%?

Your first response is likely to be 50%? You're joking. Are you mad?

But if I press harder, you will start to suggest making it out of cheaper plastics or cheaper materials or leaving bits out in order to reduce the cost.

But this is **not** Cost Reduction. This is **Cost Cutting**.

To begin to really understand **Cost Reduction** I want you to again study the item.

The item in your hand is not just a product. It is the **end result** of a very long process used to produce it. In fact, everything in your hand first came out of the ground as raw materials that were transported to another site to be refined into a base material, which was then transported to another site and processed into a base metal or plastic, which was transported again and again, from factory to factory, to convert the raw materials into parts which were then assembled into the **final product** in your hand.

Even the very simple product in your hand could take up to 10,000 individual steps from minerals in the Earth to the final product. Every one of those 10,000 steps has an individual cost associated with it.

With this understanding you are now ready to study Cost Reduction.

The Toyota View of Costs:

> *Costs do not exist to be counted.*
>
> *Costs only exist to be reduced.*

TAICHI OHNO (TOYOTA)

Competent Cost Reduction: (The correct and only interpretation using Toyota thinking)

Stop focussing on cost.

Instead focus only on the quantity, quality, and manufacturability of the actual steps in the end-to-end process.

Now let me re-ask the original question slightly differently.

Could you reduce the number of *steps* in this 10,000-step end-to-end process by **at least** 50%?

That is, could you reduce the number of steps from 10,000 to 5,000? The correct answer is, "Yes, Easily."

By eliminating 5,000 steps, we have eliminated the cost associated with every one of those 5,000 steps.

While costs are not linear, I am sure that you can now see that it is possible to reduce the cost of the product in your hand by at least 50% - without cheapening the product or reducing its quality.

In fact, more often than not, reducing steps actually improves the quality and performance of the product.

Using this same thinking, the best I have ever seen is a 93.9% cost reduction by simplifying the design to eliminate a truly massive number of steps from the process used to design, develop, and manufacture that product.

Now that you have the correct understanding, let me return to doing cost reduction wrong. i.e. cost cutting.

Cost Cutting:

Cost Cutting is not focussed upon removing steps, only reducing the cost of every part or step in the assembly. Can we make it from cheaper plastic? Can we make it from cheaper metals? Can we leave parts out? Can we get cheaper labour? Can we contract it out? Can we make it more efficiently (which will reduce a few steps only)?

But this faulty mindset will never compete with correct Competent Cost Reduction thinking.

In fact, these firms are not even playing the same game as those doing genuine Competent Cost Reduction.

Cost Cutting is also known as 'beat-up the suppliers' to get them to give a cheaper price.

So, the supplier does the same. He looks for ways to do things on the cheap. Even beats-up their own supply chain.

This is the same absolute loser attitude that comes up with the idea for 'broken biscuits' in the tearoom.

Yes, it does reduce cost by a little, but often at the cost of poor quality or poor customer satisfaction.

Cost Cutting is always part of the incompetence view of Defence.

Cost Cutting is **NEVER** Cost Reduction.

*"The most unsuccessful **four years** in the education of a cost-estimator is fifth grade arithmetic."*

~ AUGUSTINE'S LAW NUMBER VIII

Value Added:

Engineers are taught that only around 1-5% of steps in a process actually add true value to the end product.

That is, around 97% of steps are non-value adding (or 'waste' to use the Toyota vernacular).

Good engineers are taught to design products in ways that reduce non-value adding as much as possible.

But this approach still does not match competent Cost Reduction correctly done as described above.

Cost Blowouts:

Now it is easy for us to understand what causes cost blowouts.

Cost blowouts are principally caused by some need to **increase** the number of steps in the process, whether those steps are from having to redo some work, remake some parts, solve a new or unexpected problem, unresolved or left-over trade-offs, late arrival of parts, quality problems, manufacturability problems, Producibility problems, delays, a change of plans, errors in planning, errors in execution, or other missteps or mistakes.

Bigger cost blowouts simply signify a need for a very big **increase** in total steps in the process.

If we now look back at the AWD ship cost blowout, we can see that actual costs doubled the original budget. This suggests that the actual number of steps to complete the ship were around double the originally planned number of steps required to manufacture and assemble the ships.

Something as big as a ship can easily have 100 million steps to produce it. In simple terms AWD doubled that to around 200 million steps (and we can assume that 200 million steps costs around twice as much as 100 million steps).

Let's go back to the product in your hand. In simple terms the 10,000 steps to make the product in your hand became 20,000 steps to make that product. This is exactly how cost blowouts work.

Cost blowouts are always part of the incompetence view of Defence, and always welcome in the Wall Street view.

Cost blowouts are 'NORMAL' in every REACTIVE or BASKET CASE weapons development program.

Cost Padding:

The most insidious and pernicious of all activities by Defence.

It's insidious because it is hidden from everyone.

Hidden even from the Parliament and, most especially, hidden from the Taxpayer.

Cost Padding refers to any activity that artificially inflates a budget above a competent base amount to do the job.

In other words, Defence is building into its cost plan an entirely expected increase in the number of steps to complete the project beyond that required to competently build that weapon or equipment.

Cost Padding always forms part of the incompetence view of Defence.

Cost Padding is a built-in defence against entirely expected REACTIVE or BASKET CASE programs.

A Simple Cost Model:

As steps increase, costs increase. As steps decrease, costs decrease. As quality improves, costs decrease. As manufacturability improves, costs reduce. As producibility improves, costs reduce.

Reactive development methods always increase the number steps in every weapons program.

Predictive development methods always reduce the number of steps in every weapons program.

Reactive methods always increase costs. Proactive or predictive methods always reduce costs.

WBP2020 and Cost Reduction:

WBP2020 is best described as the Toyota Cost Reduction method on steroids.

Truly massive Cost Reduction is possible using this method.

The Toyota Cost Reduction method typically focuses upon reducing the number of steps in the end-to-end manufacturing process (while improving quality, manufacturability, and producibility of the product).

WBP2020 takes this thinking a lot further.

WBP2020 looks at reducing the number of steps in the entire lifecycle of the product. From concept design, through drawing in CAD, development, component manufacturing, production, assembly, commissioning trials, field operations, servicing, sustainment, upgrading, right through to dismantling and recycling at end-of life.

Removing even 20% of these steps will reduce costs by enormous amounts. That is precisely the aim.

Costs Summary:

Cost Cutting - is **NEVER** Cost Reduction.

Cost Reduction - is competent reduction of steps in an end-to-end process, often 50% less.

Cost Blowouts - are an **uncontrolled increase** in steps to complete the program, often 50% more.

Cost Padding - is **inflating** a cost budget **above** a competent base amount, often 20-50% more.

Cost Planning - is a way of thinking about how to competently do anything in the least possible number of steps.

Defence does not do Cost Reduction.

Because Defence does not know how to do Cost Reduction.

Defence only does REACTIVE programs.

So, Defence only understands Cost Padding, Cost Blowouts and *Cost Delusion*.

Thus, endeth this lesson!

An Introduction to 'Wicked' Problems.

A wicked problem is a problem that is extremely difficult or even impossible to solve because of incomplete, contradictory, and changing requirements that are often difficult to recognise. It refers to an idea or problem that cannot be fixed, or where there is no single solution to the problem.

So, in simple terms, a 'wicked' problem is an intractable or unsolvable problem.

By 2020, Defence Weapons Systems programs have become such an **intractable problem** that it easily **defeats every initiative** undertaken by current and previous Australian Governments to reform Weapons Acquisitions.

'Wicked' problems in weapons acquisition programs are an inevitable consequence of using reactive models

The entire weapons acquisition supply-chain, including the defence department and Prime contractors, have become so tunnel-vision focussed on putting band-aids upon band-aids that they have lost focus on what is important, i.e. competent weapons acquisition programs that are affordable and producible.

Defence has simply taken a system that never really worked well and by always taking the *lesser choice* ('this is as good as it gets') has turned something that worked somewhat effectively 30 years ago, firstly, into a bigger problem which has continued to deteriorate until now that system problem has finally became unsolvable.

Once this weapons acquisition problem became unsolvable, a new 'normal' mindset takes over - and everyone becomes so 'tunnel-vision' focussed to the extent that **no other possibilities exist** other than to keep **'managing' that intractable problem** forever. i.e. nothing we have done works, so it's best to just manage what we have now.

Once this mindset is established it becomes impossible to think outside the box - and an intractable problem becomes a truly 'wicked' problem that is now completely unsolvable.

'Wicked' problems then often lead to the creation of perverse incentives in an attempt to fix them.

This is the ongoing story of Defence Weapons Acquisition in Australia.

An Introduction to the 'Illusion of Depth of Knowledge'.

Much of Defence suffers from an *'illusion* of depth of knowledge'.

Just because people in Defence itself, Defence qangos, or in the Defence Primes, can TALK endlessly about Defence Weapons Development does not mean they have the actual depth of knowledge to know how to EXECUTE competent Weapons Development **outcomes**.

This *illusion* of depth of knowledge has dominated Australia's Defence's policy and programs for the past 30 years.

Defence claims the Frigates and Submarines are two of the most complex Defence programs worldwide. "The program is very ambitious and carries significant risks – particularly for cost and schedule". This is despite the opposite claim that product life cycle management tools used in naval construction have brought the "most advanced manufacturing and life cycle management capabilities to Australian industry".

(Defence really must get out of the helicopter view, onto ground level).

Likewise, many individual Defence firms (including Primes) will say they are already at, or near, World's Best Practice at Engineering methods and Development competence, when in cold hard reality they are many, many miles away from it, as evidenced by the complete failures noted in the case studies in Chapter 3.

It is this *illusion* **of depth of knowledge** that has led Defence astray - and is what directly causes repeated cost and schedule problems which in turn result in 'Projects of Concern' in Defence Weapons Development programs.

In some areas Defence **readily accepts** a *knowledge depth gap* exists. So, for example, Defence created the CDIC (Centre for Defence Industry Capability) to help to improve the operational efficiency of Defence SME's business operations to bring them up to an acceptable level.

Yet no-one will admit to any knowledge depth gap in the Prime Contractors. It is just *assumed* that all Primes already have total and complete *actual* depth of knowledge. Yet Defence continuously experiences cost and schedule problems or blowouts in projects, and delivery of on-going 'Projects of Concern' from these same Primes.

Likewise, CASG (Capability Acquisition and Sustainment Group) the Defence Bureaucracy, is not recognised as having any knowledge depth gap despite never-ending repeat of the same old problems in the programs they manage.

Defence asserts the belief that Defence Weapons programs are 'difficult' and that the type of problems that result in 'projects of concern' are simply the 'normal' outcome for all Defence Weapons Acquisition Programs.

This is simply not true. There are new World's Best Practice methods used in the commercial world which can, and do, quite easily eliminate each and every such risk.

Despite clear and persistent evidence that this '**illusion** of depth of knowledge' has existed across wide areas of Defence for many years, there is no recognition inside Defence that anything needs to change.

This is the ongoing story of Defence Weapons acquisition in Australia.

*A **Footnote:*** I am sure many readers smiled at Augustine's quote above. But I am also sure that most missed his subtlety.

...”the most unsuccessful four years”...

Augustine's dry humour suggests that most cost estimators had to repeat the fifth grade four times.

The fact that original cost estimates in defence are almost never achieved, suggest Augustine just may be right.

8

WHY DEFENCE CAN'T FIX IT

"Defence only knows what defence knows - and nothing more."

Sadly, a truism that abounds throughout Defence Acquisition programs across recent decades.

The really big question must be:

Who in Defence has the vision, authority, and especially the cojones (slang) to overhaul Weapons Acquisition?

Why can't Defence fix it?

Simple Answer: Defence does not want to fix it. That is because Defence does not believe that it has a problem.

Logical Answer: This is the way it has been for more than 20 years - so why change it now?

"This is as good as it gets"....

... *"What more do you want from us?"*

That is, *'this is as good as it gets'* - and the mediocre mindset behind it - is perfectly acceptable to Defence.

As an ex-Toyota executive, I struggle to understand how such mediocrity of leadership can exist.

But I struggle even more to understand why the military Warfighter leadership side of defence does not declare outright war on the dysfunctional weapons acquisition side of Defence. Because the resultant dysfunction from 'this is a good as it gets,' must NEVER be acceptable to the Warfighter.

Thus, if the military war-fighting leadership side of Defence has zero interest in fixing Weapons acquisition failures then no pressure will ever be brought to bear on the weapons acquisition side of defence to fix their problems.

But it does not stop there.

The Defence Weapons Acquisition side of Defence has no defined requirement for the defence prime contractors to do better than 'this is as good as it gets'. In turn, these prime contractors have no defined requirement for the firms in their supply chains to be better.

This means that the mediocre 'this is good as it gets' mentality permeates every nook and cranny of the entire end-to-end Defence Weapons Acquisition system.

But Why?

Defence Acquisition is a Monopsony. That is, Defence acts as a single Monopoly buyer of Defence Equipment that has total and absolute market power over all sellers in the supply chain.

Meanwhile, due to mergers and acquisitions over the past 30 years, the supply chain itself has today become a very narrow oligopoly of a very small number of Prime Contractors. In some cases, there is now only a single possible supplier.

This means that competitive tendering no longer works at all in Weapons Acquisition Programs.

Defence does not see any problem with being a monopsony, dealing with a small oligopoly of suppliers.

In the commercial world, precisely the same shrinking of supply chains has happened, but the response from the more astute firms is very different. Commercially, firms in a single supplier situation have moved away from using reactive methods towards proactive methods. For a small minority of very astute CEO's this has shifted even further to predictive methods. Because only predictive methods can work in a single or twin supply chain model.

But Defence still lives in the past. What once worked well for them in the past, must surely still work well today?

This is the way of the Monopsony. Arrogance makes them too slow to react to a changing world. If it reacts at all.

Debunking the Myth of the First Principles Review.

Defence is very good at defending itself from criticism. They can always point to some 'expert' review or other that has miraculously 'fixed' all of Defence's past failures. The First Principles Review is simply the latest 'crutch' that Defence relies upon to defend itself against criticisms like this book.

But the First Principles Review's ability to solve Defence's Acquisition failures is nothing but a myth. It's a myth because there were two fatal flaws in the First Principles Review.

Defence will become enraged at such outrageous blasphemy because they will argue the Review was conducted by erstwhile Defence 'experts' of high reputation with years of previous experience in Defence matters.

That is precisely the first fatal flaw problem.

The review should not have been conducted by Defence 'experts' at all. Because it is not a defence problem at all. It is a **systems** problem that is at the heart of Weapons Acquisition failures. Because it is a **systems** problem the First Principles Review should have been conducted by **'systems'** thinkers.

Those people with the expertise and deep understanding about how complex interactive feedback systems must work correctly, and how they can fail so catastrophically. It is the **system** that Defence uses that is the problem, not the specific Defence components within that system.

So, the first fatal flaw proposed a Defence solution to a SYSTEMS problem.

The second fatal flaw problem is one that cripples nearly every improvement project, restructuring, or reorganisation across every country, across every industry, and across every firm.

In our rush to implement the planned reform, most often the real hidden root-cause of the problem is overlooked. This is because in our rush we are actually only resolving the visible *symptoms* of the problem. Typically, only a cursory review of the current state is undertaken, if any review at all, because after all we are going to completely discard the current state in favour of the bold newly constructed future state plans.

In terms of time, almost all our time and effort are spent working on the future state. This seems logical as the future state is where we will be living and working once the plan is implemented.

But this thinking is fatally flawed. It leads us to ignore the constraints and limitations that are holding us back tightly into the current state right now. These constraints and limitations holding us back are the real root-cause of all our current problems today. If the planned future state solution does not specifically address each and every one of those root-cause problems, then those root-cause problems will simply move themselves across into the new future state and manifest themselves as if nothing had changed. This is exactly what happens in Defence.

I'm sure most readers will have experienced this for themselves at some time or other. A big change is made or a big reorganisation happens - but exactly the same problems reappear in the new system. The visible symptom we see may change a little, but it is the same old root-cause problem merely dressed in new clothes.

"Until the actual root causes of these persistent problems are addressed, then no amount of additional oversight, extra regulation, rearranging of organisation boxes, creation of new offices or structures, or changes to processes will help to improve the outcomes of current or future Weapons Acquisition programs. In fact, they will most likely only make things even worse".

So it was with the First Principles review. The Review did not deeply analyse the deep current systems failure modes in sufficient depth to reveal the actual root-causes of why the very same Weapons Acquisition failures continue unabated after each and every prior review process.

So, in effect, the First Principles Review simply re-arranged the deckchairs on RMS Titanic.

The Myth of the First Principles Review is simple.

Time passes, nothing changes. Or perhaps, 'Sacred Cows' remain sacred.

These two fatal flaws make a complete mockery of the supposed success of the First Principles Review.

I will describe what the First Principles Review should have looked like in chapters 12 and 17.

The Sting in the Tail

Having now been seen to have 'fixed' Defence weapons acquisition via the First Principles Review, if it does not work, then there must be some other 'mysterious new problem' or reason for ongoing Weapons Acquisition failures. Defence will simply assume the First Principles Review and its accompanying re-organisation actually worked, so they will never again revisit the actual root-cause of the problem. Instead, they will look around everywhere else to find other reasons for the continual failure of Acquisition programs.

This is a recipe for an even further deepening of the already 'wicked' problem that is Weapons Acquisition.

That is, if the First Principles Review could not fix it, then it is now an even more unsolvable problem.

And unsolvable (wicked) problems can't be fixed, only managed forever (or so Defence thinks).

If this sounds like 'this is as good as it gets', déjà vu, then you are right.

Why a Complete RESET of Weapons Acquisition Reform Will Not Happen

A complete **reset** solution like I will describe in Part two of this book will be far too high risk for Defence to consider. Over the past 20 years, and particularly the past 10 years, Defence has become far too risk averse to attempt any such bold **reset** that is so far outside of their comfort zone and knowledge space. Only a far lesser proposal which conforms to defence's now 'normal' risk profile will be acceptable to defence.

"Defence only knows what defence knows - and nothing more" limits all opportunities to do something better.

"This is as good as it gets" - and the mediocre mindset behind it - will remain perfectly acceptable to Defence.

The basic question still remains: Who in Defence has the vision, authority, and especially the cojones (slang) to overhaul Weapons Acquisition?

9

WHY POLITICIANS CAN'T FIX IT

The very first question must be:

Who among our Politicians has the vision, authority, and especially the political cojones (slang) to overhaul Weapons Acquisition in Defence?

The second question must address the problem of the political needs of the Politician.

What is the base purpose of Weapons Acquisition Programs?

The problem is the conflicting needs of our politicians means most often the needs of the Warfighter take second place to more urgent political or electoral imperatives.

The Wrong Focus.

The first political problem is that our politicians have the wrong focus behind their reasons for Weapons Acquisition Programs. Far too often for the Politicians, Weapons Acquisition for the Warfighter is subservient to Nation-Building, or Job Creation, or Innovation or some other political distraction.

This is a serious problem because Defence uses these political distractions to defend the indefensible because they provide the perfect excuse to hide their incompetence behind. By that I mean every sin of cost blowouts, time blowouts, and basket-case programs can be excused and forgiven as long as it is creating jobs or delivering Innovation or support for some State Premiers' political needs, or some Politician's electoral needs.

These alternate political priorities severely interfere with better delivery of effective Weapons Programs.

They will most certainly be used by Defence to interfere with any Weapons Acquisition Reform initiatives.

All of this allows Defence a convenient excuse for 'this is as good as it gets'.

The Sting in the Tail

Having been the instigators of the First Principles Review, our Politicians will now assume that review has worked and, because they never knew about the fatal flaws, the Politician will now assume 'job done'.

But when the next Program suffers from the same old problems, the Politician is puzzled. What now? We, the Politicians have done our job, so this must be some new problem. They can't conceive that it is simply the same old problem returned as if nothing had happened.

Having now 'fixed' Defence Weapons Acquisition via the First Principles Review there will not be a further verification review to make sure that the First Principles Review had indeed fixed the root-cause problem.

So, the Pollies now determine that Weapons Acquisition really is a 'wicked' (that is, unsolvable) problem after all which should now be left to the Bureaucrats to manage as best they can (after all it is unsolvable in their eyes).

Catch-22. 'this is as good as it gets'.

The Right Focus.

There is only ever one reason for all Weapons Acquisition Programs. To put the very best possible capability that we can afford into the hands of our Warfighters so they can successfully prosecute any war.

There are no other reasons.

But this singular purpose provides small comfort to the Politician's political and electoral needs. And this political and electoral focus by our politicians will not be changed easily.

Essentially this is saying the rot in Defence Weapons Acquisition actually starts with our politicians selfish needs.

'Shit' only ever flows downhill from there.

So, the second political problem is that perverse selfish incentives drive political behaviour.

And that will remain simply because our Politicians can't see the problem.

Given this selfish need, will any politician ever have the cojones (slang) to overhaul Weapons Acquisition?

Why a complete RESET and refocussing of Australia's Weapons Systems Acquisition System is necessary

A Defence Weapons Acquisition system with **broken DEVELOPMENT SYSTEMS capabilities** will never deliver on the Morrison Government's Defence strategy for better Defence Weapons Acquisition processes.

As we have seen previously, current Defence Weapons Acquisition Programs have some very serious root-cause failure modes that dog every existing and new defence program. These problems with Weapons Acquisitions are not getting better - but getting worse by the year. The 'classic' signs of a now unsolvable 'wicked' problem.

Why the Government needs a complete RESET to happen

Given the current Government has committed over $200b to re-arm Australia's defence forces over the next decade or two - while simultaneously creating a new sovereign industry capability to support Defence - this commitment by the current Government is such a large **nation building exercise** that it must not be allowed to fail.

But fail it will unless the 'wicked' problems are fixed by complete break-through ideas that will **stop** the old mindset dead in its tracks - and create a new opportunity to shift policy back to a right path solution.

All 'wicked' problems can only ever be resolved via a complete policy **reset** solution.

They cannot be solved by using the same mindset that created the problem.

Disruption is the only solution.

This is a critical lesson for Scott Morrison, the Morrison Government, and Defence to learn if they want long-term success in future Defence Weapons Acquisition programs in Australia.

Why a complete RESET and refocussing of Weapons Systems Acquisition will not happen

Despite the vital need, such a **reset** is unlikely to happen unless a Politician with the cojones to put aside their political imperatives and self-interest can be found to lead the type of reform needed.

Is there such a Politician in our Parliament?

Having spoken face-to-face with many, I have not seen such a person.

I doubt that such a person exists anywhere within our current political framework.

I know many will say, "I can do it." but words are cheap. Action belies words.

I say don't tell me - show me - that you can do it.

Alas, that is where it all falls down.

10

TAXPAYER PAIN

It is the Taxpayer who directly pays for the folly surrounding current Weapons System Acquisition methodologies.

What then for the Taxpayer?

Defence can't fix it.

Our Politicians can't fix it.

Perhaps 'this is as good as it gets' really is 'as good as it gets' for the Taxpayer.

Because if Defence can't fix it - and our Politicians won't fix it - then the only way left for it to get fixed is for the Taxpayer to hold our Politicians to account until they do fix it.

But that will require Taxpayers (as voters) to keep chucking out incumbent Governments until they learn to respect the will of the voters.

How likely is that to happen?

Given the myriad of other 'problems' in our economy and our society that rightly occupy the forefront of the voters or Taxpayers mind, it's probably not likely at all.

It might be important, but in the larger scheme of life, other things become more important.

Weapons Acquisition Reform is not among the Taxpayer's highest priorities.

So, the never-ending pain for the taxpayer goes on.

Is it acceptable? No.

But the people we elect, and the people we rely upon to spend our money wisely, do not in turn respect us.

So waste is bountiful, and previous attempts to stop it were fruitless.

> *"People working in the Defence sector should try to save money. There remains the possibility that it may someday be valuable again."*
>
> *~ AUGUSTINE'S (SLIGHTLY MODIFIED) LAW NUMBER LII*

But the Taxpayer is also their own worst enemy for two reasons.

1. It's not unusual to read in the public comments following an article in some newspaper or defence blog about cost blowouts to read things like 'at least the money is spent in Australia'. While such comments may have some simple logic to them, they are ultimately a complete disaster for the Taxpayer.
   Because these comments simply give the politician every excuse or justification to continue to espouse their 'Nation-Building', or Job Creation, or Innovation, or some other political distraction comfortable in the knowledge that the taxpayer will not 'beat them up' as the wasted money is 'at least being spent in Australia'.
   And of course, those political distractions will then be used by Defence to justify why nothing should change.
   So, déjà vu - these circular excuses mean that the pain for the Taxpayer will never end.

2. But even worse than 'at least the money is spent in Australia' is the even more insidious 'who cares?'
   Today we tend to see this 'who cares' response more and more as the Taxpayer becomes more and more quiescent and passive in response to the constant waste of taxpayer money.
   Because for the Taxpayer this is not just about the waste of money by Defence. For the Taxpayer, Defence is only one area of wasted money. The same cost blowouts happen all too regularly across most areas of Federal and State Government programs. Taken collectively they amount to a seemingly endless tide of waste of

Taxpayer money that no-one seems able to fix, least of all the Taxpayer.

So, it's not hard to understand that the Taxpayer suffers from cost blowout and 'wasted money' fatigue. As in, "I just don't want to hear about yet another cost blowout anymore." So, the 'who cares' response is becoming more prevalent every day as people 'tune-out' because they feel that there is nothing that they as individuals can do about it. If they vote out one mob - the new mob is no better. Time passes, nothing changes.

Which of course just allows the politicians their excuses - which Defence uses to keep the status quo.

And the Taxpayer is complicit by providing the perfect excuse for both the Politicians and Defence.

The circular excuses go on, forever undiminished.

A classic catch-22. Or perhaps checkmate.

11

WARFIGHTER PAIN

It is the Warfighter who pays the ultimate price for failed current Weapons System Acquisition methodologies

What then for the Warfighter?

If the Warfighter's own leadership can't get it fixed,

And Defence can't fix it,

Our Politicians won't fix it,

And the Taxpayer can't fix it.

What then for the Warfighter's ability to successfully prosecute a war?

Restating the Basic Principle:

The whole point of Defence acquiring Weapons Systems Programs and Equipment is so that our Warfighters can successfully prosecute any war.

Therefore, anything **at all** which impedes our Warfighters ability to successfully prosecute a war must be deemed a **complete failure** of our Defence Weapons Systems Acquisition programs and methods.

Note: there is no second place in the prosecution of war - you either win or you lose.

There is no 'honourable' draw.

To be able to successfully prosecute war our Warfighters must be given the right capability to win every time. If the equipment and capability they need is not immediately available to them - in the volume they need - when they need it - our Warfighters, and our country, will lose.

Therefore, any failure in Weapons System Acquisition is a complete and utter failure to the Warfighter.

The Warfighter Pain:

In the preceding 10 chapters we have seen a myriad of reasons that get in the way of our goal of providing the very best capability that we can afford to our Warfighters.

Time and again basket-case programs, incompetence, the Wall Street model, inept leadership, weak political will, and 'sacred cows' all combine to show us why things never change for our Warfighters. There always seems to be some other priority coming before them and their needs.

I remind the reader of the words of Sir Bernard Gray: *"If I just let defence contractor A get on with it and 'Do just please drop by and deliver the equipment' I am unlikely ever to get that equipment".*

Taken to its extreme, this suggests that eventually the Warfighter would run out of equipment to prosecute war.

This is the lot of our Warfighters.

But the Warfighter is also their own worst enemy. Never complaining. Stiff upper lip. Thankful for what they do get. Frustrated as hell that it is not better. **But silent.**

Note: this must change in future - see chapters 16-18 for how a new voice will work.

While there have been dozens of studies, reviews and failed re-organisations across defence, no one has ever sat down with the Warfighter and asked; how badly is your equipment broken? Or how does it stymie your job? Our Warfighters especially do not get asked what else we must do for them?

Ultimately the whole aim of any Weapons System program is to deliver into the hands of the Warfighter the vital capability needed to successfully prosecute war against an enemy.

This is the challenge we the People of Australia must now address on behalf of our Warfighters.

We, the people, must not shirk from the formidable responsibility of this Brutal Challenge.

This is the challenge outlined in Part 2 of this book.

A Nagging Question:

One question remains that I must answer: Don't the Military Branch Chiefs have the cojones to fix Weapons Acquisition failures?

Perhaps they do, but under the existing Defence structure they have no direct say over the Acquisitions side of Defence. Acquisitions come under the Administration side of the Department of Defence (DoD) - not the Military side of DoD. To all intents and purposes, they are entirely separate functions within DoD.

The Branch Chiefs can complain loudly, but that is all they can do.

It was interesting that the ex-Warfighters disagreed with me on this point.

The ex-Warfighters all ultimately agreed that the Military Branch Chiefs can potentially wield sufficient influence to have a significant say over **what** is acquired – IF they are astute enough at using their influence on Weapons Acquisition choices.

But they get little say about how it's acquired, or when it will turn up and that's probably not such a bad thing in the case of the former – catastrophic for the latter.

The truth is that CASG (Capability Acquisition and Sustainment Group) (and DMO [Defence Materiel Organisation] before it) are a law unto themselves, accountable to no one.

One thing that works in CASG's favour is that the tenure of their leadership tends to be longer than the tenure of the Military Branch leadership and often when their people get to senior leadership positions they will have had decades in CASG.

Meanwhile, in the Military Branch two things happen. Service Chiefs have rarely had experience in the capability development and acquisition space but, if they have any, it will have been quite minimal.

So, they confront a powerful opponent and formidable opponent in CASG.

Secondly, no Military Branch Chief will openly go against decisions already taken, especially if they themselves were on the periphery of it (e.g. when a new Branch Chief has been the deputy Branch Chief or the Chief's head of capability). To do so is to admit the organisation got it wrong and they themselves were part of that.

Not going to happen.

Even if it does, and the Military Branch Chief wins the day inside the Department, he then has to win with the Pollies, and if it's the same colour of Government it will *most certainly* never be changed.

Example: Navy's light destroyer program, the DDL – killed by Beazley in favour of FFGs off the shelf.

Super Sea Sprite helicopters, killed by Combet in favour of the latest versions of Seahawk, off the shelf.

So, the Ex-Warfighters seem to think the Military Branch chiefs do have the power. But with lesser capability and knowledge against a far more powerful and influential adversary they become effectively impotent.

12

PAST FAILURE OF REFORM

Weapons Acquisition Reform has long been a topic of conversation in almost every country that develops its own Weapons for its own military.

In Australia, no robust on-going conversation about reform of Weapons Acquisitions exists. Rather, it is more about the odd critical article, here and there, in the media or in Defence related media or blogs.

In the UK there is a more robust conversation, but again, it is more a reflection of discontent with the continual failures than it is a focussed forward-looking strategy.

In the USA it is quite different. They have a number of Government funded organisations in the Defence space whose role is to promote such discussion. This is where the Robert Gates led Acquisition Reform ideas started.

But where ever they start, they invariably fail. Weapons Acquisitions fail around the world for much the same reasons as they do here in Australia.

Australia:

There have been too many enquiries and reviews to mention them here. But the First Principles Review of 2015 was the 49th review in 42 years. On average, that is more than one review every year. While not all were about Weapons Acquisition, they were all about the continued failings of

Defence programs. All have failed to deliver reform. All had similar results. Not much changed. The 'sacred cows' remain undisturbed.

These 49 reforms have failed to make any significant improvement in Weapons Acquisition results.

The incompetence model always remains intact. The Wall Street model always remains intact.

The First Principles Review was just the latest. Chapter 7 debunked the Myth of the First Principles Review by pointing out two fatal flaws in the process. Actually, there was a third fatal failure mode brought about from the implementation of that Review's recommendations. Because the review itself was flawed, then the recommendations from that review were also flawed. The solution implemented was flawed because it was fixing the wrong problem. Instead of fixing the real problem it merely re-arranged the existing deckchairs on the Titanic. Some of them in ways that have made the original problem worse.

That review saw the old DMO (Defence Materiel Organisation) morph into CASG (Capability Acquisition and Sustainment Group). In this process, the previously semi-independent CDG (Capability Development Group) (the voice of the Warfighter) was subsumed into CASG or back to the Military services. But CASG is dominated by the Bureaucracy. Rightly so, because their role is to protect the nation's legislative interests around Legal, Fiduciary, and Probity issues in spending so much taxpayer money.

But, by subsuming CDG's voice of the Warfighter role beneath the legal, fiduciary and probity issues, they in effect gave total control of weapons development to the purchasing and legal department and removed the engineering department. This put the bureaucracy role on the customer (Warfighter) weapons development line.

They should have done almost the exact opposite. A new role (Military Capability Group) beyond even CDG's Military's role should have been established and put on the Customer (Warfighter) line. They should have then expanded this new role to include an Engineering Capability Group.

This would have made a customer line of Warfighter -> MCG -> ECG -> Prime contractors. The CASG or DMO role should have been retained but moved **off the Customer line**. Competent Weapons Development requires a Warfighter to Engineering conversation along the customer line, not a business development or sales conversation.

More on this solution in Chapter 17 where we discuss solutions to the Defence swamp.

In conclusion, these 49 reviews in 42 years prove something else. That is, that Reform cannot succeed if done from within the Defence System. Reform will only ever succeed if the existing Defence swamp is disrupted.

Without external disruption, the Incompetence model and the Wall Street model will continue to dominate.

And the 'sacred cows' will remain sacred.

The Losers?

The Taxpayer, and the Warfighter.

As always.

USA:

Similarly, in the USA there have been quite a number of enquiries and Reviews. Again not much changed as evidenced by the comments in the excerpt from the Pentagon's Office of the Director, Operational Test & Evaluation (DOTE) FY13 report in the 'last word' at the end of Chapter Three. It said, *"Looking at three decades of reliability assessment, the report concludes that the reliability of DoD systems has not significantly improved over time"* under three different policy settings.

The last major attempt at Reform was under Robert Gates described previously in Chapter Seven. While Gates' intent (or rather those under him) was indeed noble, ultimately they failed. They failed because their work never translated into the planned legislation update to WSARA2009. (Weapon Systems Acquisition Reform Act 2009). While WSARA2009 was a first step based on moving from reactive to proactive methods, it has largely failed to be enforced, or been ignored in the return to traditional reactive methods of the Defence swamp.

But there was another ultimately more serious failure mode in the reform that Gates implemented in 2009. Gates never legislated how to **execute** the WSARA 2009 legislation. So, the US DoD was able to mostly ignore the legislation and continue using the same old execution methods that had failed again and again.

The basic idea of Gates was sound and provides an insight into what path the next stage of reform should take.

Gates intent was to bring a program down to two final participants. Gates would then fund both to create an entirely new type of Concept Design based on the Munro pilot program work, which created a very robust and detailed business case with predicted Affordability and Producibility indexes that define the predicted Cost, Predicted Quality, Predicted timeline, Predicted Sustainment Readiness Level, Predicted Producibility Index (technical capability) and predicted Confidence Index (how well it will work in the hands of the Warfighter).

Gates then had to simply chose the Prime with the best Affordability and Producibility to be awarded the contract to design and build the weapons system. The beauty of this way was that the firm which chose a Predictive method would always win over any firm which continued to use any other method. Reactive methods would *always* lose.

For Gates, this was the ultimate 'NO BULLSHIT' version of Weapons Acquisition. Marketing hype and bullshit could never survive the intense scrutiny of the planned brutally effective Predictive Acquisition model.

Although Gates ultimately failed to get this new model legislated, his work does provide a solid foundation for the next attempt at generating Weapons Acquisition Reform around the world.

Even where there was only a single source supplier, the Toyota model provided insight and guidance. The exact same method could be applied but, instead of a competitive competition, a 'Design to Target' model would apply. That is, Gates would set an affordability target, and the sole supplier had to deliver a concept design and business case that met every target set for that program before a contract could be approved.

Does it work? This is precisely what happened during the pilot and proof of concept developments. A target was set, which the concept design and business case met, and then the actual program result delivered into the hands of the Warfighter two or three years later was within 5% of that original business case target.

If, for example, we could apply these pilot and proof of concept results to the AWD program, the Australian Government would have avoided spending any of the extra $5b that was eventually wasted. In other words, AWD would have come in for less than the original budget plus contingency set for that program.

I sometimes wonder if Gates knew how close he was to success. Success, at least, in creating a new Weapons Acquisition model that worked. Getting it legislated would have been a whole different problem.

Those with the most to lose, those who rely so heavily on the organisational incompetence model and the Wall Street model, would have fought tooth and nail to defeat it. Why? Because they would not survive the disruption that would follow on from the application of such legislation.

Hidden behind the wall of organisational incompetence, the laziness created by the Wall Street model, and their reactive mindset, they would not now be nimble enough to survive the mad scramble to comply with the new legislation. For them, it is only about their own survival not about better Weapons Acquisition.

Self-interest always wins

The Losers?

The Taxpayer and the Warfighter.

As always.

Footnote: The final special Defence version of Gates model does not exist. The pilot and proof of concept work was prototyped within MDS using basic modifications and workarounds. The program was killed before the final special Defence version was ever formalised. Likewise, the special Defence version later proposed for Australia's SEA 1000 Submarine program was never completed. What does exist today is a commercial world version which does not require anything like the robustness developed for the special Defence version. e.g. Defence specific SRL (Sustainment Readiness Level) is not required in the commercial world. Nor is the more specific and complex work required to establish the cost per hour to fly a modern high-tech fighter plane (or ship or vehicle).

However even using the commercial world version would be far better than anything Defence uses now.

But the basic knowledge behind how to restart and recreate the special defence version still exists. It is simply waiting for someone, somewhere, with the cojones to restart Weapons Acquisition reform.

The full Defence version could be fully developed inside 12 - 18 months if required.

13

CONCLUDING THE NEED FOR REFORM

Back on the first page of Part 1 - The Defence Swamp, I wrote:

SWAMP: Current Usage: Government Bureaucracy of dysfunction, waste & incompetence, **wholly unfit** for purpose, **wholly unfit** for office.

A question for the reader: Does this statement adequately describe the outcomes of Chapters 1 thru 12?

I think a reasonable person would conclude that current Defence Weapons Acquisition methods are a Government Bureaucracy of dysfunction, waste and incompetence, wholly unfit for purpose, and wholly unfit for office.

Certainly, a great many of those who have experienced it for themselves seem to think so. I have found no shortage of people in Defence Weapons Acquisition that have directly told me so. And the facts and actual outcomes described previously speak for themselves.

Sadly, we used to be much better at this than we are now.

That of course raises the next question: What can we do about it?

Complete disruption is the only way forward.

Without external disruption, the Incompetence model and the Wall Street model will continue to dominate.

The 'sacred cows' in Defence will remain forever 'sacred'.

The Losers? The Taxpayer and the Warfighter. As always.

But an underlying question remains: Who has the cojones to overhaul Weapons Acquisition?

Time and again, Defence methods have been subject to review after review with little, if any, improvement.

Perverse incentives abound across Weapons Acquisition that reinforce bad behaviour and bad outcomes.

Competence of leadership to deliver better Acquisition ways are non-existent in Defence.

The needs of the Warfighter are almost an afterthought in Weapons Acquisition.

The need to spend Taxpayers' hard-earned money wisely is not respected.

The vital need for reform is not even a topic for conversation in Defence.

Affordability and producibility are not part of the Defence lexicon.

Protecting 'Sacred Cows' has become the norm.

'This is as good as it gets' prevails.

Just as it has for the past 25 years.

Time passes, nothing changes.

The Swamp lives on. Undisturbed.

But the nagging question remains: Who has the cojones to overhaul Weapons Acquisition?

I will conclude the case for Weapons Acquisition reform with these brilliant (modified) words of Augustine:

> *"The more time Defence spends talking about what they have been doing, the less time they have to spend doing what they have been talking about. Eventually, they spend more and more time talking about less and less until finally they spend all their time talking about nothing."*

> *~ AUGUSTINE'S (SLIGHTLY MODIFIED) LAW NUMBER XLVIII:*

These final words provide a good introduction into the Blueprint for much needed weapons acquisition reform.

Chapters 1 thru 12 outlined the need for reform.

It is now time to mark the end of the gloom, despair, and darkness of the first 12 chapters.

We first saw a faint spark of a new light in chapter 4, which became a little brighter still in Chapter 7.

It is now time to farewell the darkness forever.

It's time to look forward towards a far better, much brighter future.

A future where .... the swamp is **drained!**

II

HOW TO DRAIN THE DEFENCE SWAMP - A BLUEPRINT FOR REFORM

A BOLD AND DARING WEAPONS ACQUISITION REFORM PROGRAM

Welcome into the light.

A new and bright future where Affordability and Producibility is respected.

A new and bright future where the Taxpayer's hard-earned money is respected.

A new and bright future where the Warfighter as the centre of the universe is respected.

A new and bright future freed altogether from the past.

> *When you look at the world, do you see the same problems and solutions as everyone else sees ....*
>
> *... or do you see a different possibility?*

Chapters 14 thru 21 outline a Blueprint for a new Weapons Acquisition Reform methodology.

But true Weapons Acquisition Reform will require a robust solution to the Defence Swamp.

SWAMP: Current Usage: Government Bureaucracy of dysfunction, waste & incompetence, wholly unfit for purpose, wholly unfit for office.

A Solution to the Swamp:

Government Bureaucracy of World's Best Practice, competent, functional, highly effective, wholly fit for purpose, and wholly fit for office.

The following seven chapters will outline both an idea for a new Weapons Acquisition Reform methodology, and a practical solution to the swamp.

Proposing a new weapons acquisition methodology is easy.

But nothing will change unless our Government passes a new legislation framework that will force a change of thinking upon the entire end-to-end Weapons Acquisition delivery system.

That is the challenge accepted over the next 7 chapters.

A Précis:

A robust solution requires resolution of every problem described in Part 1 - The Defence Swamp.

If Defence Weapons Acquisition is a Swamp - then the Swamp must be drained.

If obfuscation and 'muddy' facts are the problem - then facts or 'near facts' are the solution.

If Alternate realities are the problem - then fact-based realities are the solution.

If reactive methods are the problem - then predictive methods are the solution

If 'management by stress' is the problem - then SYSTEMS thinking is the solution

If Organisational Incompetence is the problem - then high-performance Knowledge Centres is the solution.

If Wall Street is the problem - then affordability combined with supplier profitability is the solution.

If Defence can't fix it, then we must develop a new way and compel them to use it.

If the Politician can't fix it, then we must create a new way that makes it all too easy for the Politician to execute.

If the Taxpayer can't fix it, then we must fix it for them.

If the Warfighter can't fix it, then Weapons Acquisition must be rebuilt with them as the centre of attention.

If the status quo is the problem - then disruption is the answer.

These are the challenges accepted over the next 7 chapters.

> *You can't solve problems with the same thinking that caused them.*
>
> ~ ALBERT EINSTEIN

But an inconvenient niggling question remains:

Who has the political cojones (slang) to Reform Weapons Acquisition in Defence?

14

WEAPONS ACQUISITION REFORM - AN OVERVIEW

Before a better Weapons Acquisition method can be discussed, first we must decide on the standards needed to define success. We already know what failure looks like. But what does success look like?

Success is always based on creating and establishing Superior **Systems**.

The Success Australia Needs:

A more robust Program Management and Risk Control method that will meet and deliver the following:

Aims:

1. A singular, deliberate focus on delivering absolute best capability into the hands of the Warfighter.
2. **Affordability** and **Producibility** as paramount mandatory considerations in every acquisition program.
3. Zero 'Projects of Concern' in Weapons Systems Acquisition programs.
4. Zero possibility of any repeat of Collins or AWD problems in Weapons Systems Acquisition programs.

Ideal Success outcomes:

1. 30% less time to design, develop and introduce new Weapons Systems Acquisition programs.
2. 30% lower cost of acquisition of all Weapons Systems Acquisition programs.
3. 30% lower cost of sustainment and Thru Life Support/Ownership Costs in Weapons Systems programs.
4. 70% improvement in quality and reliability of Weapons Systems equipment.
5. 80% improvement in producibility of Weapons Systems equipment.
6. 90% lower program risk in Weapons Systems Acquisition programs.
7. 100% of required Weapons System Capability in the hands of the Warfighter, and on time.

Note: when I say 30% lower cost, I mean 30% lower than original budgets (not cost padded budgets).

The Question for Defence is – How Can They Achieve These Aims and Outcomes?

Because nothing they are capable of today can deliver these expected results.

Therefore, simple changes or improvements to current practices will not work.

Only a full root and branch change to the way Defence thinks about Weapons Acquisition can deliver the results that the Taxpayer and Warfighter demand from those who will deliver Weapons System capability in future.

That will require a whole new mindset about how Defence thinks about Weapons Acquisition.

That will require a whole new leadership model for Weapons Acquisition.

That will require a clean-out of all of the 'dead-wood' that exists today.

That will require someone inside Defence with the cojones to **reform** Weapons Acquisition.

That will require someone with the political cojones (slang) to drive Weapons Acquisition **reform** in Defence backed up by expert **systems** and an expert group of people orchestrating everything.

That is the truly **brutal** challenge we face in our goal to Reform Weapons Acquisition in 2020.

We, the people, must not shirk from the responsibility of this brutal challenge.

'This is as good as it gets' thinking must die, before we can go forward.

The great challenge lies before us.

There must be no 'sacred cows'.

The true path forward is not difficult.

Others have already shown us a new way forward.

The more astute CEO's and firms from the commercial world who have already blazed their own trail to reform, have shown us a new path to follow.

These more astute CEO's have already shifted away from the old reactive models that Defence clings to.

These more astute CEO's already know that only highly predictive methods work in the modern world.

These more astute CEO's have already sought out people to help them create the future.

We simply need to tap into the World's Best Practice thinking that these astute CEO's have already started.

The next four Chapters will outline what our future would look like.

15

INTRODUCING PREDICTIVE THINKING

The best way to design and develop any product is not to design and develop a product at all.

The best way to design and develop any product is to first design and develop a **system** for how to design and develop a product. Then the product will virtually design and develop itself, without problems.

SYSTEMS thinking:

One of the best **systems** thinkers was Dr Edwards W. Deming. Deming is more popularly known as a Quality Guru. But in fact, he was actually more interested in how Systems Design causes variation (problems) within the system. So, while most of his work is associated with fixing quality problems in the factory, he believed the root-cause of those problems being felt in factory was in fact being caused in the design of the product or system.

Deming's mantra at that time was:

Every business is made up of systems.

All systems suffer from variation (problems).

So, most of Deming's work became associated with reducing variation (meaning problems) in the factory.

Note: what follows is not Deming's work on TQM or TQC. It is based on Deming's belief that if you DESIGN the SYSTEM right - then there is little need for the use of TQM or TQC tools to later fix 'problems'.

Deming's systems thinking is also the genesis for both the Toyota Development System (TDS) and the Munro Development System (MDS). Deming worked for many years with Toyota on quality systems, not just in the factory, but also in design. Sandy Munro worked with Deming for a number of years in Ford.

Sandy Munro tells the story of how he came to leave Ford and setup his own business in Deming's image.

After a particularly frustrating meeting with Ford's then president, Deming turned to Sandy Munro and said,

> *"Son, these people will never get quality. You should leave and setup your own business to fix quality at the source, in the design."*

Having decided that Deming's brutal assessment and direct insult to Ford's worldwide President was probably a career limiting move for Sandy within Ford, he took Deming's advice and left Ford.

It seems God does indeed move in mysterious ways.

The Munro Development System was born with Deming as its genesis.

So now the Deming/Munro mantra became:

> *Every business is made up of systems,*
>
> *All systems suffer from variation (problems),*
>
> *All variation (problems) start in the design of the system*

Therefore, we must design the **system** first, before designing the product.

Success is always based on creating and establishing Superior **Systems**.

In designing such a system, we must use a set of robust rules and tools that will eliminate, or reduce, all variation (problems) **before** we design the actual equipment (plane, truck, tank, ship or submarine).

Otherwise we will simply design-in variation (problems) in every product or program. Unless of course luck intervenes. But luck is never a controllable or repeatable process.

As Deming always said; "The root-causes of a problem must always be fixed before any problem can be permanently resolved." Fixing symptoms of a problem is just an exercise in frustration.

Problems Thinking:

Taking Deming's thinking further means understanding the type of variation (problem) and the impact or scale of the variation (problem) on any system.

Toyota developed a set of simple metaphors for this purpose:

Boulders: Problems so big they would completely disrupt and stop dead any development until fixed.

Rocks: Problems still big enough they will disrupt and stop any development until fixed.

Stones: Problems still big enough that they will interfere or delay any development until fixed.

Pebbles: Lower level problems so numerous that they interfere with any smooth development process.

Sand: Minor problems that the system can handle on a day-to-day basis without too much disruption.

Powder: Best thought of as the normal background 'noise' in any system.

Now we can bring these two ideas together.

TDS requires that **all** boulder, rock and stone level trade-off problems MUST be resolved before we 'cut steel'.

WBP2020 requires all boulder, rock, stone and as many pebble trade-off level problems as possible **must** be resolved before we draw anything on paper or in CAD.

We can't eliminate every single problem because we are human, and humans do not have perfect knowledge, so we cannot expect to eliminate every single problem before it exists. The point is to eliminate every problem of a level of variation that will delay or interfere with a smooth design, development, and production of that product.

Let me demonstrate with a practical example.

In 2007, Toyota and a competitor were developing a new model on similar timelines. The critical program timeline gateway is N-6 in Toyota speak. In your speak, that means 6 months before the planned production start date.

In 2007, Toyota passed through the N-6 gateway with 23 open problems that they did not at that time have a final solution to. One of those was a stone level problem, the rest were either pebble or sand level problems. That means that on average each Toyota engineer had 0.3 of a problem to solve in the next 6 months. So, the Toyota engineers are focused on start-up quality, manufacturability, production preparation etc.

Meanwhile the competitor passed through their N-6 gateway with around 1900 open problems. A mix of boulders, rocks, stones, pebbles, and sand level problems. On average every engineer had between 30 and 50 problems to fix within 6 months. What are the engineers focused on? Fire fighting, and even more fire fighting of problems. No-one is focussed on helping the factory with quality preparation, manufacturability, production preparation etc.

This is a pure 'Management by Stress' system. The humans are stretched to and beyond breaking point.

In Toyota, the stress is put on the system by the design of the **system**. So, the humans can do their job properly.

Which of the two examples do you think came in on budgeted cost and with the best customer satisfaction?

Defence is like the competitor. Entirely Reactive. The only difference is that for the competitor the start of production date never moves. In Defence the start of production date is completely flexible and can easily move out by months, even years. Despite moving the production dates out, they will still experience many problems in production and in the hands of the Warfighter. Reactive mediocrity rules.

Predictive Thinking:

Predictive thinking simply takes Deming's systems thinking to a new level not possible in his day. While Toyota developed its original Proactive model based around Deming's 1970's thinking, the technology to move to Predictive thinking has only become available in the last 15 years. Toyota today is somewhere between Proactive and Predictive. It is the Munro Development System (MDS) that has far surpassed what Deming thought possible.

As we saw back in Chapter 6, the catalyst to move towards more predictive thinking in Weapons Acquisition came from the US Office of the Secretary of Defence under Robert Gates. Gates, if you recall, began asking if MDS could be modified to bring a major decision point back before milestone A (drawn in CAD) i.e. to create a fully predictive model.

While that model eventually failed to be implemented in Defence, it has had success in the commercial world.

The basic logic is sound.

Predictive thinking works.

The Solution:

Any Proactive method will be far better than any Reactive or Basket-case method

Any Predictive method will be far better than any Proactive, Reactive or Basket-case method.

The challenge is no longer to create predictive thinking.

The challenge is to get firms to use predictive thinking.

16

PROCESS REFORM

Designing and developing a product is actually very simple.

What makes it so hard is the level of disruption and interference caused in reactive or basket-case programs.

Predictive Systems where disruption and interference either don't happen at all, or only at very low levels, makes product design and development quite easy (it's now a rewarding process - instead of management by stress).

Introducing WBP2020:

World Best Practice 2020 (WBP2020) is a convergence of ideas over the past 10 years into a single methodology designed to deliver more affordable products or services with much higher quality and lower costs.

WBP2020 is a hybrid of the Toyota Development System (TDS), the Munro Development System (MDS), and my own methods developed from my own learnings and accumulated knowledge from working in truly high-performance knowledge centres and teaching that capability to others.

The whole purpose of WBP2020 is the deconstruction of complexity into much simpler outcomes.

This book is not designed as a tutorial in precisely how WBP2020 works. Rather, it is about conveying an understanding of a robust solution that eliminates all reactive and basket-case Weapons developments.

I don't expect the reader to comprehend the intricacy or elegance of this model. Merely to grasp that it is a highly predictive methodology designed specifically to substantially reduce the cost of products, programs or services, while simultaneously shortening the time taken to develop and introduce those products, programs or services.

At its core is **Affordability** and **Producibility**.

There is a critical reason why a new method like WBP2020 must be used in all future Weapons Acquisition programs. That reason is: up to 90% of all cost, quality, time, functionality, manufacturability, producibility, sustainment and affordability problems and trade-offs are 'locked-in' the instant the capability (equipment) is drawn on paper or in CAD. This is because the program development timeline is tight and will not allow extra time for any major redesigns. At least not without endless delays in delivering the needed capability to the Warfighter.

If the design locks-in up to 90% of total lifecycle costs, then it's pointless trying to make design changes after the design is locked down as it must inevitably blowout costs and blowout the timeline.

Therefore, any changes at all that are made for any reason after it is drawn in CAD will always lead to increased time and increased cost. As we saw back in Chapter 6, under Cost Reduction anything which increases the number of steps beyond that required for a competent design and development will always increase costs and always increase time taken. This is precisely the nemesis of today's Weapons Programs.

So, any reform of Weapons Acquisition must mandate any predictive method that will resolve 90% of problems and trade-offs before anything is drawn on paper or in CAD. The aim of any reform must be to eliminate 90% of design and development problems before they ever exist on paper or in CAD. Therefore ~80% of all weapons Acquisition effort *must* take place *before* it is drawn in CAD (i.e. do it Right First Time). Then, once robust control is established, we *must maintain total control* through-out the entire end-to-end Development cycle.

This is precisely what WBP2020 is designed to do.

Does this mean that everyone is locked in to use WBP2020 only? No. Just as Robert Gates never legislated MDS in WSARA-2009. What Gates legislated in 2009, and planned to do in 2013, was to legislate the deliverable outcomes from using a highly predictive methodology. His prototype and proof of concept was MDS, but his planned legislation would allow for <u>any</u> predictive method that delivered the same operational results as MDS.

In my case I have been immersed in TDS for 33 years and using MDS on and off for 19 years. So, for me WBP2020 is just simple, natural thinking. This book is talking about WBP2020 because I am writing it. But there are others out there working on how to create Predictive systems. MDS has the immediate advantage because it was the pilot for Gates' ideas.

Now all this extra work upfront sounds like it takes longer, yet you say it shortens timelines. How come?

Let's go back to Chapter 6 under cost reduction and the number of steps. Something as big as a ship might take 100 million individual steps or tasks. So, if I add an extra 3 million steps to put a predictive design method in place, the number of total steps will increase to 103 million steps. But we must look at the entire end-to-end system to see what happens next. That is, because we will eliminate a huge number of trade-offs and problems from the EMD phase (engineering and manufacturing development) we will reduce that part of the development by say 30 million steps. The net effect is a reduction from 100 million steps to 73 million steps (100+3-30). This is exactly how a predictive method can **spend a lot more time upfront yet reduce the total program timeline by 20-40%.**

While many of the terms used in WBP2020 are familiar to Defence, the content within those terms is vastly different. For example:

Contestability:

This is way beyond anything Defence normally does. It is the elegance and thoroughness of Toyota, and the brutal strength and the relentless questioning and inquisitiveness of Munro.

Conops:

A concept of operations way more detailed than anything Defence uses. A Toyota level of depth and understanding.

Concept Design:

Like no other concept design in Defence. Concept design is actually a poor description of what this process does and the outcomes it delivers. **The depth to which it goes is extraordinary**.

Business Case:

Does not adequately describe what this process delivers. It is so robust that in the commercial world **it can quite accurately predict what a program will cost per hour to fly a plane before any physical drawings exist of that plane**.

> *For those following Munro's current teardown of the Tesla S, and Tesla Y models, using the reverse engineering version of MDS, Elon Musk has tweeted that Munro's costs are accurate. Car companies have also reported MDS costs are within 1% of their actual costs.*
>
> **For those who use LEAN:**
>
> *Think of this process as a value stream map of extremely fine infinite detail on steroids. Detail down to the individual 100 million steps in the ship example. Each individual step has cost, quality, time, manufacturability and sigma level (predicting likely failures) attached to it. All before drawings exist.*

This is the power of Predictive thinking developed a long way beyond anything Deming could have ever imagined.

But this also illustrates another possibility for the future of Weapons Programs. While the core focus is on how to design new weapons programs, equipment or products right the first time, the beauty of both WBP2020 and MDS is that they work just as effectively on remediation programs.

That is, where a weapon system or product already exists and has problems, it can be put through a right-second-time activity that will reverse engineer the equipment or product to its infinite detail. Which will reveal precisely where all the problems lie. Then we simply design forward again eliminating the problems as we go.

This version is the ideal candidate to be used for example in the Collins lifecycle extension program. This program requires the existing Collins submarine with all of its problems to be redesigned to extend its operating life by around 15 years until the new French submarines are battle-ready in the hands of the Warfighters.

The remediation version described above can analyse the existing boat plus add in the new design elements inside a single contiguous development process.

Program Management and Control Methods

The above process will require a new method for how all future programs are managed such that it will eliminate all of the existing problems and deficiencies in Defence's current methods.

WBP2020 uses a modified version of Toyota's Obeya and Gateway methods and Munro's Wall process, combined with my own methods and modifications to suit Defence specific requirements.

Toyota's Obeya process and Gateway Control methods are like nothing Defence uses. In my 23 Years Toyota introduced over 800 New Vehicle programs - every one of them was delivered without slipping a single day.

Munro's Wall process is an integral part of their Predictive thinking processes.

WBP2020 Example Outlines

On the next two pages I have included a simple outline of two ways that a highly Predictive method like WBP2020 can be used.

A: as a right-first-time method for the design and development of new products or services.

B: as a remediation tool to fix problematic weapons programs and equipment that already **exist.**

Example A: as a right-first-time method for the Design and development of new products or services.

A: WBP2020 Outline: Development of New Weapons, Products or Programs

Product Imperative: Right First Time (because coming off 2nd best is **never** an option for the Warfighter).

Supply Imperative: Create the best possible Capability Solution that Australia can afford.

Program Imperative: Start with the end in mind (100% Battle-capable in the hands of the Warfighter).

Key: How to help Defence *acquire* capability, and how to help Defence *deliver* that capability.

Development System Imperative: a robust front-end problem-eliminating methodology.

1. Warfighter Focus:
- Contestability (+ Robust Conops)
- Producibility
- Sustainment

Note 1: Clarify in great detail the <u>USER</u> needs, <u>Client</u> Needs, and operating environment the Capability must work in when in the hands of the Warfighter.

2. Needed Capability:
- What to Acquire
- How to Achieve
- What to Deliver

3. Technology:
- Technology Space Maps
- Technology Transfer
- Technology Support

4. Risk Modelling:
- MRL, TRL, SRL
- Producibility Index
- Confidence Index

5. Innovation Pipeline:
- Research Effectively
- Develop Effectively
- Execute Effectively

6. Affordability:
- "Must Cost" Targets
- Price Must Never Trade-off Volumes
- Cost Blowouts are Not Accepted

7. Concept Design: (Concurrent Engineering)
- DFA, DFM, DFSS, DFSC, DFS
- Program Viability Modelling
- Highly Robust Business Case

Note 2: Items 1 thru 7 must be completed prior to CAD.
Note 3: Robust CAD Design only comes AFTER we know what to Design and that it will work and be 100% battle-ready in the hands of the Warfighter

8. Technical Excellence:
- Design Methodology (CAD, Solvers, etc))
- Development Methodology
- Toyota Program Control Methods (Gateways, Obeya)

9. Manufacturing:
- Production
- Supply Chain
- Kaizen

Design & Development Imperative: Early Control is Vital to Success

WBP2020 Purpose: Create Warfighter Advantage while delivering Producibility and Affordability.

You will notice the enormous amount of effort in items 1 thru 7 that must precede the start of the actual physical design of the product or weapon system (item 8).

Notes:
MRL = Manufacturing Readiness Level
TRL = Technology Readiness Level
SRL = Sustainment Readiness Level
DFA = Design for Assembly
DFM = Design for Manufacturability
DFSS = Design for Six Sigma
DFSC = Design for Supply Chain
DFS = Design for Sustainment

A Simple Example: When presenting from stage, I would often use an extremely simple example based on the above predictive method that reduced the number of steps to make it from 146 to 8 and which reduced the total cost by 61%. Yet when directly challenged no-one in the audience could tell which was the old part and which was the new part.

(Sorry, no Defence examples because no one's Defence secrets can be given out via this book).

Note: WBP2020 requires every step in this entire end-to-end process to be executed competently.

You are not doing WBP2020 if you only do some of these steps or even if you partly do most steps. WBP2020 success requires a full and total commitment to competently execute the whole process.

Example B: as a Remediation tool to fix problematic weapons programs and equipment that already **exist.**

B: WBP2020 Outline: Remediation of Existing Weapons, Products or Programs

Product Imperative: Right Second Time. (Fixing problematic weapons programs that already **exist**).

Supply Imperative: Remediate the Capability Solution to be fully problem free for the Warfighter.

Program Imperative: Start with the end in mind. (100% Battle-capable in the hands of the Warfighter).

Key: How to help **Recover** Warfighter capability, and how to help **Improve** that Capability.

Development System Imperative: A robust method that finds and eliminates all root-cause problems.

1. User (Warfighter) Focus:
- Contestability (grasp the actual problem)
- Producibility Needs
- Sustainment Needs
Review in great detail the USER needs, Client Needs, and operating environment Problems the Capability is experiencing in the hands of the Warfighter

2. Reverse Engineering:
- What is the Problem?
- Where is the Problem?
- How Serious is the Problem?

3. Review Business Case:
- Viability
- Risks
- Legal

4. Quality:
- Imperatives
- New Quality Target
- Quality Solution

5. Cost: (Affordability)
- Imperatives
- New Cost Target
- Cost Solution

6. Functionality:
- Imperatives
- New Functionality Target
- Functionality Solution

7. Re-Design: (Concurrent Engineering)
- New DFA, DFM, DFSS, DFSC, DFS
- New Producibility
- New Serviceability (Sustainment)
Note 2: Items 1 thru 7 must be completed prior to CAD and before re-manufacture begins.

Design & Development Imperative: Remediate Problem Capability

WBP2020 Purpose: Remediate Cost, Quality, Functionality to eliminate ALL Problems

You will notice this remediation method uses WBP2020 in reverse.

Typically, it will create either a field fix for the problem or a whole upgrade of the weapon capability or sustainment.

There is no equipment problem, or operational or sustainment problem that this method cannot fix.

It is quite often used to reduce the cost of a product, or to reduce the cost of sustainment of existing equipment.

A Defence Example: McDonnell Douglas/Boeing C-17.

An earlier version of Munro's MDS methodology was used during a partial remediation program during the development of the C-17. The resultant remediation road map when implemented has resulted in C-17 now being regarded as one of the most successful Airforce Programs when in operational use in the hands of the Warfighter.

- **Part Reduction** = 46,135
- **Operating Steps Reduction** = 273,008
- **Time (hrs) Reduction** = 5,550
- **$ Saving per Aircraft** = $1,371,001

Raytheon: 18 Year Problem... Solved In 4 Months - 50% Reduction in Weight - 72% Reduction in Cost

Astute Submarine: Cost Reduction Program on Boats 4-6 (delivered targeted Cost Reduction amount)

Note: WBP2020 remediation requires every step in this entire end-to-end process to be executed competently.

You are not doing WBP2020 remediation if you only do some of these steps - or partly do most steps.

WBP2020 remediation success requires a full and total commitment to competently execute the whole process.

17

STRUCTURAL REFORM

Introduction of an entirely new and predictive method reform like WBP2020 will have no effect unless the structure in which it is used is also reformed.

The basic question that must be asked is: What is the best possible structure or system that will maximise the benefits from WBP2020 in Weapons Acquisition that will deliver the best possible results to our Warfighters.

Like Product Development, it is about designing a new Superior **System** first.

Like WBP2020, we don't design the Management System first. We need to develop the Customer Line first.

Leadership and the Management System must never be on the direct Customer Line.

This means we need to focus upon what to include on the Customer Line. The Customer Line must be optimised first, and then work out how the other necessary bits can be added in the least intrusive way possible.

Step 1: In Defence Weapons Acquisition the correct Customer line must start with the Warfighter at one end and the Equipment suppliers at the other end. So, the absolute simplest Customer Line looks like this:

Warfighter ◄─────► Equipment Supplier

Note: for many this graphic will seem the wrong way around. i.e. many will think that It should have the suppliers on the left and Warfighter on the right. But this is wrong thinking. In the real world the ONLY thing that matters is the battle-ready weapon actually in the hands of the Warfighter. That is, we must start with the end in mind. Once the desired end-point capability is understood, we then work backwards to ask if we have the right suppliers, with the right capability, to be able to deliver 100% of that battle-ready capability to the Warfighter. To use the Toyota vernacular - everything else is a 'bullshit fact' - it does not exist in reality.

This process effectively reverses the current method used by Defence.

Step 2: Now we need to add in a very simple project control method between these two. That is, we create the simplest **ideal** customer line that will work without problems. It might look like this:

Warfighter ◄─► M-CKC ◄─► E-CKC ◄─► Equipment Supplier

This has added M-CKC (Military Capability Knowledge Centre) to focus the voice of the Customer, and E-CKC (Engineering Capability Knowledge Centre) interpreting that Customer requirement into required technical equipment to deliver the Warfighters desired weapons capability. So, we have now created 2 separate sub-systems. One in the military side and one in the equipment side. This is the minimum viable system required to create a fully functioning and ideal Customer Line.

Step 3: Now we must add in a Leadership and Management system in the **least intrusive way possible**. Ideally, we don't diminish the Customer Line at all, or to the absolute minimum necessary. But note that Leadership and Management must **never** be on the direct Customer Line. It is always a support or servant function. Never a critical element of the Customer Line.

Step 4: Now we must add in the minimum necessary bureaucracy in terms of the Government's mandatory need for legal, fiduciary and probity oversight in the least possible intrusive way. The bureaucracy must NEVER, EVER, be on the Customer Line. The bureaucracy must **never** diminish the Customer Line.

If we have done all of that right, we will have created a structure that will best exploit the full potential of WBP2020.

Note: none of these roles are the same as they are today. **This is not the old CDG in new clothes.** *This is a whole new approach to a totally new Defence role. M-CKC and E-CKC are actually Knowledge Centres of high human capability about how to execute high performance product development. They are not fixed departments.*

They are functional roles that can call in whatever knowledge is required to produce the right outcome, even if that extra human knowledge must come from outside of Defence. **They are not reliant on the Primes for knowledge.**

The First Principles Review did not follow this path. Instead, it did the reverse and tried to reorganise conflicting priorities in the opposite order. i.e. it tried to optimise the three different components of the system and then attempted to put them together like different parts of 3 different jigsaw puzzles. It 'sort-of' fits... and if you get a big enough hammer it can all be made to fit, but in ways that totally disrupt the effective flow of the Customer Line.

The result? A newly modified incompetence model. Looks different. Sounds different. Works much the same.

This now raises the next question. What to do about the existing leadership, management and structure?

There is only one answer. The old ways are totally incompatible with this new Customer Line Reform model. Not because WBP2020 and this new Customer Line model are hard to learn, they are simple, but it's because the old reactive and basket-case ways are too hard to unlearn and leave behind.

There is simply not enough deep dissatisfaction within the current leadership and management systems. I don't mean they are not unhappy at all the stress and problems, it's just that the dissatisfaction level is so low that it does not drive a strong enough desire for complete root and branch disruption. In a sense the status quo is comfortable. This is largely because the existing leadership and management have grown so accustomed to 'managing' the truly 'wicked problems' in Defence on a daily basis that they can't see a better solution when it comes along.

These are tough decisions. But there can be no 'sacred cows' in the reform of our dysfunctional Weapons Acquisition swamp. Preservation of 'sacred cows' is why every previous reform has failed.

A Practical Example: The Customer Line in action (a simplified narrative).

Note: this is a completely made up example, lest I give away someone else's secrets.

The process is non-linear - but I will explain it linearly for ease of comprehension.

The desired Weapon is a small Missile system.

Facilitator. Me.

Timeline. 5 working days

The Scene:

In another country (Defence does not require this method to be used in Australia).

A very big windowless room (with lots of wall-space to put things on).

Lots of whiteboards and flip chart paper (for visual thinking and retention of ideas and thought process).

Seven Military personnel of various rank (foot-soldier through to General).

Five Suppliers of key components (Warhead, Rocket system, Fuel system, Guidance system, Body Structure).

The Process:

Day 1 begins with questioning of two foot-soldiers from different battle groups about how the weapon is used, what's wrong with existing weapons, what needs to change in future weapons. Then we move on to a Logistics low-level officer to explain how the weapon system is packaged, stored and delivered to the Warfighter in a battle zone. This is followed by a unit commander who explains the military combat aims. Then a mid-level Strategy and Tactics officer begins the conversation around the military purpose of the Weapon system. This is followed by the military program manager for this type of Weapon system. Finally, the General will explain how this Weapon system fits in to his overall battle plans.

As the day progresses the questioning becomes more and more intense. Questioning everything. Leaving nothing to chance. For example, two key operational targets may come out of this: the missile needs to go further, and it needs to be capable of going 'thru the window' 100% of the time with no collateral civilian damage.

Day 2 begins by switching to the supply chain and each supplier in turn. This involves intense questioning of their competence as a supplier and their capability to deliver what the Warfighter needs to fight a war. Typically, this stage will include a simple Toyota QFD (Quality Function Deployment) chart. This compares the voice of the Warfighter with the suppliers engineering requirements.

(Note: in 30 years of doing this no-one has ever been able to complete this task beyond about 20% - such is the poor understanding of what we actually know vs what we think we know - typically it then takes between 4 weeks and 1 year to complete a robust full QFD chart)

Day 3 will switch back to the Military. Now the contestability gets very intense. Every statement made on Day 1 is challenged in a very robust exchange. It is quite common that large parts of what they think they need will fail to pass muster under this barrage of contestability. But that is the whole point. To eliminate all errors of thinking.

Day 4 switches back to the suppliers. Each supplier in turn now faces a barrage of robust question about their ability to design, develop and produce the battle-ready weapon system as described by the military. It is not unusual for the conversation sometime during the day to go like this:

"Mr Warhead Supplier, will the warhead you initially proposed do what the soldiers want?"

"Well, no."

"Do you have a warhead that will do what the soldiers want?"

"Of course, we do."

"Then why did you not use it?"

"Because the original specification (specs) did not call for it."

"Mr Rocket Supplier, will the rocket you proposed do the extra distance the soldiers want?"

"NO!"

"Do you have a rocket motor that will do what the soldiers wants?"

"Of course, we do."

"So why did you not use it?"

"Because the specifications did not call for it.

"Mr Guidance Supplier, you heard the soldiers say they want it to go through the window 100% of the time. Will the proposed guidance system do that?"

"Well, no."

"Do you have a guidance system that will do that?"

"Of course, we do!"

"Then why didn't you use it?"

"Because the specifications did not call for it."

About this time in the process, the General typically starts moaning about 'Homer Simpson' and begins softly muttering expletives and starts gently banging his head on the desk.

To the uninitiated this all sounds quite ridiculous, but having done so many of these over the years I am no longer surprised at how often the specifications and the needs of the Warfighter or Customer do not match.

I think one of the biggest learnings for all of the participants in this process is to grasp how poorly we specify to the supply chain exactly what the Warfighter actually needs to win a battle. Then we wonder why the supply chain makes such a mess of developing the Weapon system as shown in the case studies in Chapter 3.

Day 5 then begins by sorting all of the concepts, ideas, flip chart sheets, etc around the room in a storyboard. Starting with how the Weapon System will be designed and developed. How it will be produced. How it will be tested and commissioned. How it will be stored and transported. How it will be used in battle. That is, a complete Conops (concept of operations) map is produced of the entire end-to-end Weapon System.

At the completion of this 5-day process I always ask these same two questions of the participants.

Do you now understand why Weapons Acquisition is usually such a basket-case of failure modes?

Without the knowledge you have now gained through this process, how on earth have you designed and developed all of your previous Weapons systems?

I never receive an answer because everyone is too embarrassed to provide one.

The Output:

The output of these 5 days is a 90% completed contestability and concept of operations (Conops) map, fully ready to start the concept design and business case stage.

The Point:

The whole point of using the above process is so that everyone ends up with a CRYSTAL-CLEAR understanding of the final battle-ready weapon in the hands of the Warfighter. The five key suppliers have a **crystal-clear** understanding of exactly what is required of them to competently design, develop and deliver the full 100% of that desired battle-ready capability into the hands of the Warfighter, on time, without fail.

Only with this **crystal-clear** knowledge can we now design what we **must** design, not what we **can** design.

The Why:

Because there is no second-best outcome for the Warfighter in any battle.

The Future:

In future, the facilitation role will be done by E-CKC. Not just this facilitation, but they will also facilitate the concept design and development of the business case. Once the project moves into actual development at the supplier level, E-CKC shifts to a program management role starting when the suppliers start drawing in CAD through development, production, commissioning and only ends when the weapon is declared 100% battle-ready in the hands of the Warfighter.

Now think about the sheer intellectual and operational horsepower that will accumulate within E-CKC over time.

Imagine the level of accumulated human capability and competence had we started this process 20 years ago.

In the short-term E-CKC is about developing a weapon. But in the long-term E-CKC is about accumulating a truly massive knowledge base about technology and weapons capability that is unsurpassable by any defence prime.

Notes: I think anyone can see that 'bullshit facts', all marketing hype, and any/all wrong thinking will be exposed for the frauds they are when subject to such intense scrutiny by this process.

The marketing hype described in the addendum to the JSF case study and the 'French Folly' cannot survive the robust methods, contestability and proactive thinking as described in this example

18

FUNCTIONAL REFORM

Having first put in place a highly predictive thinking methodology and then a new robust Customer Line model, the remaining reform is functionality, or how it will work.

The major functional change in this reform is that **the new Customer Line must become a high-performance knowledge centre of high human competence and excellence in weapons system development.**

WBP2020 is both a way of thinking and a method to create a high-performance knowledge centre of excellence in high human competence and capability in Weapons system development.

WBP2020 is a capability not a brand.

The core purpose of the military sub-system of Warfighter to M-CKC is to create a high-performance knowledge centre of competency and excellence in the capability necessary to successfully prosecute war.

The core purpose of the equipment sub-system of E-CKC to supply chain is to create a high-performance knowledge centre of competency and excellence in the capability necessary to successfully design, develop and deliver the necessary technology so the Warfighter can successfully prosecute war.

This combination now creates a third critical sub-system of M-CKC + E-CKC which is actually the key success factor knowledge centre in the whole process. This sub-system's core purpose is to build a level of competence **such that M-CKC + E-CKC no longer have to rely upon primes for any critical knowledge.** That is, their combined technical competence at least matches that of the primes or, preferably, far exceeds that of the primes.

The Military sub-system, because of its nature, is a difficult knowledge centre to create because of the continual rotation of key staff on a 2-3-year rotation cycle back to their military branch. This will require a change in M-CKC to include some key senior direct operational staff who never rotate. Some of those I suggest should be Ex-Toyota SYSTEMS thinkers. This is to aid in the retention of the high-performance knowledge within this role and across its interrelationship with E-CKC. M-CKC is also not simply a representative of the three branches of Navy, Army, Airforce. It is a cross branch role focussed directly on the inter-related Warfighter needs, not the branch needs.

In this model the key operational staff in E-CKC never rotate. This is done so that E-CKC can build a robust ongoing engineering to engineering conversation with the supply chain. Some of those key staff will be ex-military engineers to maintain the strong bond with M-CKC. Some I would suggest are recruited from highly competent and experienced ex-Toyota people who are highly skilled in TDS, Obeya and the Toyota Gateway methodology.

Armed with these core skills, E-CKC now becomes the centre for program management and control of all programs. Armed with WBP2020 type tools described previously as the special Defence version, they can now easily manage and control every program to eliminate all of the problems of the past.

None of the above is entirely new. All of it comes from work in this field by firms like Toyota, Munro & Associates, and other commercial world best practice firms.

What works very well for them has simply been translated into the Defence environment.

These structures work very well for others.

They can work well for Defence with some effort.

Political Decision Making

The original model planned for Robert Gates provides guidance.

Currently our Politicians have to make very expensive and very high-impact decisions about Weapons Acquisitions based solely upon the unproven *promises* of capability and performance made by the Defence Prime.

This will stop. In future the tender will first down select to two potential suppliers. Those two potential suppliers will then be paid to complete a concept design with program viability modelling and a highly robust business case as described in example A: items 1 to 7, in Chapter 16 (i.e. the WBP2020 model).

Both potential suppliers would be mandated to create a very robust and detailed business case with predicted affordability and producibility indexes that define the Predicted Cost, Predicted Quality, Predicted Timeline, Predicted Sustainment Readiness Level, Predicted Producibility Index (technical capability) and predicted Confidence Index (how well it will work in the hands of the Warfighter) for submission to the Government.

Now the Weapons Acquisition decision for the Politician is easy. They will simply select the best outcome for the Warfighter from the two options. All marketing hype and political influence are eliminated from consideration.

Some may question the need to pay an unsuccessful supplier. But the bigger picture provides the answer. Is it not better to pay say $20 million to the unsuccessful supplier in order to *guarantee* that $2 billion is not wasted?

But it also not completely wasted. The unsuccessful supplier has learned from the experience and grown their design and development capabilities such that they will be a better potential supplier for other future weapons programs. In this sense we are paying to build better knowledge centres within the Defence supply chain.

DELIVERING RESULTS

Delivering results will consider three aspects.

A. **Expected results from reform**
B. **What to do about current programs?**
C. **The evolved Collins 3.0 as a backup option**

A: Expected Results from Reform

The pilot and proof of concept from Gates' model and other work from the commercial world provide direct evidentiary proof of what is now possible.

1. Better affordability, producibility and manufacturability in every program is possible.
2. The total acquisition cost of every program can feasibly be reduced by 20-50%.
3. Programs can be delivered battle-ready into the hands of the Warfighter within 5% of the original affordability target established during the concept design phase. This means that cost blowouts can be eliminated entirely from all future Weapons Acquisition Programs.
4. Programs can be delivered into the hands of the Warfighter in 20-40% shorter timelines. In very big programs this can translate

into **delivery of capability to the Warfighter many years earlier than is possible today.**

5. The cost, for example, of smart shells and small missiles can often be reduced by 30-50%. This allows for an increased volume to be purchased of say 30-50% more shells or missiles for the same total cost as the current purchase. This allows the extra volume to be used to increase training in the use of the Weapon system.

6. The costs to sustain (service, maintain, upgrade) the program over its life can be slashed considerably. By designing for sustainment during the concept design an SRL is established and then controlled all of the way through the acquisition phase until it is in the hands of the Warfighter. Lifetime sustainment costs can be reduced by 20-50% using this method.

7. By designing for sustainment, the operational availability of the equipment (or uptime) in the hands of the Warfighter is greatly increased. This is because the equipment has been designed to be serviced and maintained in the shortest possible time, in the simplest possible way, so that the capability is out of the hands of the Warfighter for the least possible time.

8. Because every program could be developed for example with a cost per hour to fly a plane, the predicted total lifecycle costs of every program can be known upfront before any program is approved. This will greatly aid the Government and Defence to best husband taxpayer money to decide what equipment should be produced when, and in what quantities, to best suit the needs of our Warfighters' need to be able to successfully prosecute any war. Governments would never need to allow for cost blowouts.

These results fully justify the use of the disruption model described in the preceding chapters.

These results demonstrate that the old ways must die a quick death.

These results show a new method that is wholly fit for purpose.

These results will deliver a 'draining of the defence swamp'.

Can these results all be achieved from day one? No. It will take some time to build the competency around these new ways of working. It typically takes any firm around 3 cycles to attain a solid base competency. But even the first cycle always results in far better outcomes than anything delivered using old reactive thinking and methods.

While the above describe direct results, there are a multitude of indirect results that will flow from this reform.

1. It will create a far more competent supply chain for all future Weapons Programs.
2. It will create high-performance Knowledge Centres of capability that can spill-over into other industries.
3. It will make the job of the Politician much easier because it is fact based, not marketing hype based.
4. It will reduce the budgetary requirements of Defence spending.

Success is about creating and establishing Superior SYSTEMS for Weapons Acquisition programs. Those Superior SYSTEMS described in Chapters 15 to 19 then deliver the outcomes described above.

Those Superior **Systems** enable the Swamp to be drained.

Exports

In a future where affordability and producibility is the norm, Australia should expect that we can export much of the same equipment that we build for ourselves to our global partners, or design, develop and build affordable new military products and technologies that are specifically required by other countries.

While we do export some Defence equipment today these are more down to those very few firms with really good internal design and development capabilities. It is true that we export some equipment that is not so affordable or producible, but that usually relies upon the Australian taxpayer to partially subsidise the export by paying for the problematic or 'basket-case' development programs in the first place.

Imagine now a future where WBP2020 methods are the norm for every firm in the Defence supply chain. Every firm is now using predictive methods which will deliver far more affordable and far more producible programs.

In this scenario there is no reason why Australia cannot expect a massive increase in exports of Defence equipment to our partners around the world.

Re-Equipping in Wartime

A vital issue, never discussed in Defence, is how we would quickly build any replacement equipment during a war to replaces any losses that would happen in any major conflict.

While the 'continuous build' option sounds good in peacetime it will lead to a complete disaster in Wartime.

During the Second World War, the USA was building replacement ships in a little over 40 weeks by late 1943. In 2020 we aim to take 5-7 years to build ships. How could we possibly replace these ships in the event of any war?

By spinning out work over many years to avoid the so-called 'valley of death' for lack of work, we are establishing a 'lazy' process that does not suit the quick replacement needs of ships and submarines during any war.

The problem with Defence's thinking is that they will simply say - we will speed up the process. But speeding up a 'reactive' or 'basket-case' program will simply create a complete disaster (see appendix C: Superiority).

The biggest risk is in fact moving too fast when the system itself is incapable of supporting that speed.

> *"The only thing more costly than stretching the schedule of an established project is accelerating it, which is itself the most costly action known to man."*
>
> *~ AUGUSTINE'S LAW NUMBER XXIV*

While WBP2020 delivers around 30% shorter program timelines, this would not of itself address the very much shorter timelines needed to build replacement ships during any war. But the advantage of using such predictive methods like WBP2020 means the risk of accelerating replacement shipbuilding would be much lower simply because there is far less to go wrong when the process is sped up.

WSARA 2021 Delivered Benefits

- It will completely eliminate old style cost and time blowouts.
- Defence can simply view two indices and know if the whole program is under control.

- Defence can have absolute confidence that planned Warfighter outcomes are being delivered.
- There is upfront understanding of all ramifications of design and development on cost & time outcomes.
- Manufacturing is included in the design and development process.
- The voice of sustainment will be heard early during the concept design.
- Finance can feel totally secure in the numbers.
- Defence can promise the Government with confidence.

Other Benefits:

Prime Contractors - Now have a new method to give Defence what it wants, when it wants it, at the price it wants to pay, with far fewer problems. While still allowing the supply-chain to be profitable.

Defence - Gets affordable, producible equipment, on cost, on time, and with the Warfighter outcomes it wants.

Warfighter - gets vital 'battle-ready' equipment they want with the functionality they need, when they need it.

Defence Minister - Imagine if you will - never having to announce another 'Project of Concern' to the media maelstrom that is baying for the blood of someone for yet another dysfunctional Defence program, or from the all too regular cost or time blowouts that are the nemesis of Defence today.

CASG Program Manager - When you are to be held personally accountable for the whole project's ultimate delivered success, you want to be damn sure the system you use delivers the best results for you every time.

B: What to Do About Current Programs?

(1) New Programs

Future Frigate: Hunter Class Frigates - BAE Systems

Given this is an Australian program under direct Australian control and provided by a company with many other interests in a number of other programs means it is highly likely that **this program could be made subject to new weapons acquisition reform methods while it is still in the early stages.**

Future Submarine: Attack Class - Short-fin Barracuda - Naval Group

This Development program is already moving into the basket-case stage less than 2 years after contract signing.

Given the existing major difficulties with getting the French to comply with the original contract terms related to design and build in Australia vs what the French want built in France, what is the likelihood the French would agree to any reform of the Weapons acquisition to use a new predictive methodology like WBP? Nil?

Naval Group is a single program French government-owned supplier that Australia has no influence over.

Additionally, there is a substantial penalty cost today of a little over $400m to cancel the French program.

But these issues must be considered in light of the terrible disaster for Australia should we continue to go forward only to find in 5 years' time that we have no viable submarine to build, because the program is a complete disaster.

To start again in the mid-2020's would imply an early to mid 2040's battle-ready alternative at very best.

Is this program 'too big to fail'?

Would a government faced with a complete disaster still go ahead because there are no other options left?

Australia has a very serious decision to make in 2021 concerning what to do about this program.

The first question: is this program even recoverable?

If not, what other option is there for Australia?

The original Japanese boat is not suitable. Although it is 4000t, it is a double hull, so it is really only a 3100t equivalent. It would need a completely new development program to start again from scratch.

The original German design was a new development based on a much smaller submarine so we would have to start that design again from scratch.

Evolved Collins: The only **viable** short-term recoverable option is the Evolved Collins as presented to the Prime Minister by a group of ex-submariners in late 2018. Their Evolved Collins proposal is based on work that SAAB has done to develop their A26 oceanic ER (Extended

Range) - a derivative of the original Collins family - into a genuine option for replacement upon the predicted failure of the French Folly.

Given the extensive work that has already been done, they have estimated that the first Evolved Collins could be in the water and battle-ready at least 5 years before any French submarine on today's timeline would be.

Given that this would be a new program contract **it could be made conditional upon using all of the new weapons acquisition reform methods.** (See also C: The Evolved Collins 3.0 as a backup option).

Collins: Collins lifecycle extension program - TBA (Collins 1.5?)

This is the ideal candidate to be used for a first pilot for Weapons Acquisition Reform in Australia. This program requires the existing Collins submarine, with all of its problems, to be redesigned to extend its operating life by around 15 years until the new French submarines are battle-ready in the hands of the Warfighters.

The remediation version described previously can analyse the existing boat, plus add in the required new lifecycle extension design elements inside a robust single contiguous development process.

(2) Remediation of Programs
JSF: Lockheed Martin U.S.
Given this is a US program there is no chance of them adopting Australian reform methods.

P.O.C's: (Projects of Concern) can be remediated successfully using the remediation version described previously to reverse engineer the equipment to uncover the root-causes of its problems and then design out those problems.

Sustainment: Weapons programs with very high sustainment problems or excessive costs can all be remediated back to a lower cost base, or a problem free future, using the remediation version described previously.

C: The Evolved Collins 3.0 as a backup option:

I have used the term Evolved Collins 3.0 to distinguish this proposed option from all previous Evolved Collins proposals put forward over the past 10 years (including that of late 2018).

I put forward this idea as a proposed full back up to the French submarine program.

This would provide our Government and our Nation with a second option that does not exist today. This second option would be at low cost and would not impede the primary French option in any way.

This approach also makes the most sense given the now inescapable requirement to extend the lives of the 6 Collins class submarines by 10-15 years.

Given the extreme RISK to Australia's sovereign Defence from any full or significant failure of the French Submarine program, I believe Australia would be best served to develop this Evolved Collins 3.0 idea at least to the completion of the newly proposed Concept Design and Business Case stage in Chapters 16 to 18.

Australia would then have around 3 more years of breathing space to see if the French program can deliver what it promises or whether it continues into a full or substantial basket-case predicted result.

If, in mid 2023, we find that the French program is looking more like a disaster, then a switch to the Evolved Collins 3.0 option could be taken without any delay to the delivery of new submarines into the hands of our Warfighters.

The only downside to implementing this Evolved Collins 3.0 backup plan is, of course, around $100 million that would be sunk into the early stage design work.

The question for all Australians, and especially the taxpayers, is this: Is this $100m backup plan spent today a good insurance policy against substantial failure of what is now looking like a potential $80-$90 billion disaster?

In terms of Risk Management, a 0.125% insurance policy seems very cheap and good value for money. But it is still $100m that we should not have needed to spend.

While I have no particular direct knowledge of the specific program details as presented to the Prime Minister by the ex-submariners who put forward their original proposal, we do have to trust that these ex-Warfighters would not put forward their version for an Evolved Collins Idea if they were not sure that it was a completely viable alternative to the French submarine.

My Evolved Collins 3.0 idea takes their strategy and equipment proposal and adds the new Weapons Acquisition methodologies as described in Chapters 15 to 18.

In this way, we can virtually guarantee that an Evolved Collins 3.0 design option would more than equal anything that the French program has proposed, while delivering actual submarines into the hands of our Warfighters around 3 years earlier than the French, and for a total program cost saving of around $30 billion of Taxpayers hard-earned money.

Additionally, this viable backup option would put pressure on the French program to pick up their game or lose the entire program.

If, in 2023, Australia did decide to switch to the Evolved Collins 3.0, the financial loss will be the $450 million spent on the French design program. But we are contractually committed to pay that anyway even if we cancelled today.

So, if we must pay the $450 Million either way, then we should allow the French design to be completed in 2023 and then compare their business case and costings against the Evolved Collins 3.0 business case and costings.

My recommendation to our Government and Parliament would be to fund the front-end design of an Evolved Collins 3.0 to the completion of the business case as described in Chapter 16 so that our country has a viable backup plan that can GUARANTEE that we can ward off any complete disaster if the French program fails.

We must never forget the purpose of the submarine.

The issue is not the money.

The issue is how will our Warfighters best fight and win a war.

A failed or failing French submarine is not the best solution.

An effective submarine, battle-ready in the hands of our Warfighters is the best solution.

20

THE STING IN THE TAIL: A CHAPTER OF WARNINGS

The Germ Theory of Management

The Germ Theory of Management describes a situation whereby major change or transformation takes a generation (around 20 years) before it becomes the accepted norm.

In the days before germs were discovered, all surgery was very high-risk for a patient until the existing practices around good personal hygiene and hand washing by doctors and sterilisation of instruments became the norm. Not knowing about germs, the old doctors would use the same instruments on multiple patients and transfer blood from patient to patient on their hands. Far too often this had fatal consequences for the patient.

But this blindingly obvious, much-needed, change in behaviour had to wait for the old doctors, with their established habits of poor hygiene, to retire or die and be replaced by the younger doctors who practised good hygiene and sterilisation as they had learned in their training.

Translating that to business and management, the germ theory of management suggests that even blindingly obvious advantages to Defence as described in Chapters 16 to 19 will take a full generation (~20 years) to be adopted if things are left to run the normal course of events.

If the changes proposed in this book are to be implemented quickly, then we cannot wait for the current generation who control Defence Weapons Acquisitions to retire or die before change happens.

Transformation as proposed in this book will only happen quickly by disrupting the status quo.

Disruption will only happen through new legislation which mandates the needed changes.

New Ways and New Technologies

New ways and new technologies are not the simple answer to deliver better outcomes.

Defence will tell you that they have fixed all the previous problems using new ways.

But, agile methods + reactive thinking does not result in predictive outcomes, just faster basket-case programs.

Artificial Intelligence + reactive thinking does not result in predictive outcomes, just perfected reaction.

Industry 4.0 + reactive thinking does not make smart outcomes, just faster, more accurate rubbish.

Digital shipyard + reactive thinking is not the 'magical' solution, just automated organisational incompetence.

New skilling programs and new colleges + reactive thinking simply delivers the same old outcomes.

New ways and new technologies will only deliver better outcomes with proactive or predictive thinking.

Without a complete change in thinking upfront it is most likely that new ways and new technologies will simply deliver more-of-the-same outcomes. Yes, it will be faster, yes it will be more accurate, yes it will look sexier.

But garbage in, still equals garbage out.

If This Stuff is so Good, Why Isn't Everyone Already Using It?

Three Main Reasons:

1. Most CEO's are not even remotely interested in being World Competitive. Most CEO's only want to beat their nearest competitor, or the guy next door, not be best in class. This means most CEO's are not even looking for World's Best Practice in anything. They seem to think modern business leadership is hard enough without trying to be the best at anything. But the truth about leadership is the other way around. It's actually far easier to run a truly high-performance business than it is to run a more-of-the-same average type business. As a reference, a few years ago a bunch of high-experience ex-Toyota people got together talk about how well LEAN methods had penetrated manufacturing around the world. Even the most optimistic thought that only around 2-3% of firms had done a really good job of adopting and implementing LEAN methods. The more experienced thought that it was less than 1% that had been really successful. Some thought that it was more like 0.1%. Australia is just worse than most. Many Australian benchmarks fall well short of more competitive nations.
2. Many CEO's today work to the Wall Street view of business. So, it's more about managing the business based on money, than it is about being a better business. They are far more interested in activities that create short-term money outcomes than they are about delivering future long-term organic business growth outcomes.
3. True World's Best Practice is, these days, only sought out by contrarian CEO's. Those CEO's who quite deliberately set out on a different path to everyone else. CEO's with this mindset are always actively seeking out new ideas and new Innovations around how to be much better than today. These contrarian style CEO's are always prepared to disrupt their existing business if that will lead to a better future business. For these CEO's money is an outcome that allows for more ideas and new methods and not a business driver.

But the combination of these three reasons also creates a fourth reason. Today, there are so few Contrarian Consultants left out there who know how to competently **execute** the type of different thinking and methods as I have described in Chapters 16-19. Almost all consultants today have followed the lead of the CEO into making businesses more efficient at extracting money from existing operations.

Defence, Defence Supply Chain, Defence Consulting Firms, Academia.

This book proposes a disruptive innovation to Defence, its supply chain, Defence consulting, and academia.

As such they will react much like the taxi Industry around the world has reacted to Uber - not by adapting to the new reality - but by trying desperately to cling to the old existing monopoly status quo that has served them so well, and has so successfully prevailed for them for the past 30 years.

Given that Defence, Defence Primes and their suppliers, or Defence Consulting firms, have 10, 20, or even 30 years invested in the old ways and its 'sacred cows', will they throw away all those years and all that acquired knowledge? Will they unlearn everything they currently know - to learn a completely new way of thinking, or will they defend their long-term investment in the status quo and its 'sacred cows'?

Defence, and Defence Consultants, will tell you this book is just the old ways in a different form.

It is not.

While this may seem critical of consultants, it is not. It is not about the consultants at all.

It is about whether our Government or Defence business leaders can use the disruptive framework and mental models described in this book to create enduring success in future Weapons Systems Acquisition programs.

With or without consultants.

This book advocates a completely different way forward for future Defence Acquisition Programs.

Defence and the Defence Consulting industry will continue to advocate that they follow the old way that has delivered to them the promised land, but delivered basket-case outcomes for the Taxpayer and the Warfighter.

"Hiring consultants to conduct studies can be an excellent means of turning problems into gold, your problems, into their gold."

~ AUGUSTINE'S LAW NUMBER XXXII

Those who cling to the ways that have failed in the past cannot help to create the required depth of understanding and knowledge that will be needed if we are to implement Weapons Systems Acquisition Reform in 2020/21.

Beware of the "Instant Expert"

A word of warning to those Government Leaders or Business CEO's who may decide to try and implement the new frameworks described throughout this book.

I think one of the key failings of all consulting has been the rise of the 'Instant Expert', the person who read something in a book and who now proclaims loudly that they are a master expert in the knowledge.

One of my most severe criticisms of LEAN is that the 30 year veteran of Toyota who has learned the vast depths of understanding TPS the hard way, by struggle, sweat, and doing - and the person who read it in a book, and gained a very shallow understanding - seem to have equal standing for delivering consulting capability in the LEAN world.

Given that this is a completely new Weapons Systems Acquisition framework to readers of this book - almost all the content is also completely new to Defence, Defence Primes, the Defence Consulting profession and Academia.

But that will not stop them hanging out a shingle the day after reading this book or starting a 'marketing' campaign that says, "I am a certified expert in WBP2020 and the new core competency model of Weapons Acquisition."

They harbour no apprehension that 'having read it in a book' today - and calling themselves an 'Instant Expert' at the content is a dangerous role to claim, especially if they knew absolutely zero about it the day before (that's just the way modern marketing works).

I am not saying you shouldn't use a consultant. I am saying that you should go into any such activity with your eyes wide open. Understand that you will be **paying** most consultants, who start with a very shallow understanding, to learn the framework while experimenting on your business - and you will pay them very handsomely while they learn.

That is probably not a recipe for success for most.

Awareness derived from a book does not ever equal competence.

This warning does not apply to competent consultants like Munro & Associates and those few others around the world who have been developing better methods or working on WBP2020 type models for more than 20 years.

I am simply saying, when approached by any consultant - that you ask that consultant to **show you,** by example, how deeply they know and understand any framework or how long they have been using their parts of the knowledge within that framework. Don't just allow them to tell you with 'marketing speak' they know the framework.

"Don't tell me - show me!" must be your response.

The 'Instant Expert' cannot help you to create the necessary depth of understanding that you will need if you are to change Defence Weapons System Acquisition methods into a future framework that is 'wholly fit for purpose'.

The Final Sting in the Tail

No-one should underestimate how difficult it will be to change the status quo in Defence Weapons Acquisition and the deep determination to retain the 'sacred cows' that exist today. This resistance will be fierce and persistent despite the apparent ease of implementing the solution as described in Chapters 15 to 19.

This Chapter of warnings should not be ignored by anyone.

Next Steps for You

You have now read this Book.

You have read about the failures.

You have read about the solution.

What is the next step for you?

Why should you care?

Perhaps you are appalled at the failures described in Chapters 1 to 12.

Perhaps you are appalled at the obscene utter waste of Taxpayer money.

But what will YOU do about it?

That is a question only you the reader can answer.

Perhaps you are more interested in the solutions in Chapters 14 to 19.

Perhaps they point to a potential solution to your own problems.

What then will be your next steps?

If you want help, then you will need to start looking for predictive thinking consultants.

Typically, they will be 'Contrarian' Consultants.

Those whose job it is to NOT just listen to what you THINK you need, and then simply regurgitate it back to you, but instead those who can tell you what you really need to know.

Normal consulting is reactive thinking.

Consultants using reactive thinking by will not create any vastly better future than you have right now using existing reactive thinking and reactive methods.

Where do you find Contrarian Consultants?

Why, right under your nose.

They have always been there.

It's just that you can't see them because you were never looking for them before.

You will only ever find what you are looking for.

So, you will need to change what you are looking for, in order to find the right help.

"Always remember. Your focus determines your reality."

~ GEORGE LUCAS

21

LEGISLATING THE WEAPON SYSTEMS ACQUISITION REFORM ACT 2021 (WSARA-2021)

Proposing a new weapons acquisition methodology like WBP2020 is easy.

But nothing will change unless our Government passes a new legislation framework that will force a change of thinking upon the entire end-to-end Weapons Acquisition delivery system.

Therefore, I am proposing the Weapon Systems Acquisition Reform Act 2021 (WSARA-2021).

Learning from Robert Gates' mistakes, I am also proposing that the WSARA-2021 Legislation must include how to EXECUTE the Legislation as described in Chapters 16, 17, and 18 so that Defence cannot get around implementing the WSARA-2021 legislation by continuing to use their existing failed methods.

Today in the aftermath of the great shutdown of the economy from Covid-19, the Australian Government must husband every available dollar and put it to its most efficient use in the recovery of Australia.

The Australian people and our Government can no longer afford, nor tolerate, a Defence Weapons Systems Acquisition methodology that is seen as far too costly, overly complex, and far too slow to respond to an ever-changing world we now face or which results in massive cost overruns and long program delays.

Only by introducing a newly legislated methodology like WSARA-2021 will Weapons Acquisition improve.

Only a new methodology which will deliver much better decision making, much earlier, and with much lower risk, which will work right-first time, and right every-time it is needed, can help to achieve the aims and outcomes of a 'now fixed' Weapons Acquisition methodology.

Australia needs a far more reliable and better Weapons Acquisition program management and risk control system for all future programs, with needed Warfighter capability, at a cost that is affordable to the Taxpayer.

What Australia can achieve by implementing this highly effective WSARA-2021 Legislated program.

1. Reduce the program risk on all new developments, technology and the complete programs.
2. Create a new level of product design & development capability and knowledge up-front.
3. Dramatically improve the functionality and integration of new technologies.
4. Dramatically improve product quality, reliability and functionality.
5. Dramatically reduce both acquisition & whole of life operating costs.
6. Achieve 10-50% 'whole of life' reduction of program costs.
7. Achieve 10-30% reductions in development program timing.
8. Comprehensively remediate persistent problems in existing equipment and programs.
9. Ensure all programs are achieved on-time and on-budget without trade-offs or remediation.
10. Deliver better desired 'battle-ready' Warfighter capability when they need it.

What MUST be Legislated:

1. The WSARA-2021 legislation itself.
2. How to EXECUTE the legislation (as described CH 16 to 18).
3. The need for a robust business case before CAD and using verifiable independent costing.

4. SRL (Sustainment Readiness Level) as a part of the business case and maintained through the program.
5. Full Defence ownership of IP for all taxpayer funded developments and technology.
6. Rights to use supply chain developed IP to ensure the Military has the right to fix MRO issues in the field.
7. A Toyota style 'gateway' project management control system to prevent all timeline slippage.

A Call to Action.

Prime Minister, given the critical financial pressures that face your government in funding the vast array of needed future Defence capability which, when combined with the need for more affordable acquisition of programs and sustainment outcomes, you must adopt new methodologies that can deliver this much needed program capability in ways that are much more affordable, and much less problematical to your government than Defence's current methodologies.

By sticking to the 'tried-and-true' current way that Defence currently uses to manage its existing procurement programs and sustainment efforts then you will not, and cannot, ever achieve your own goals as Prime Minister.

Only a new form of affordability and producibility modeling with associated risk maturity modeling like WSARA-2021 that represents a whole new industry best practice can deliver the type of outcomes that you will need to achieve in future Defence procurement programs over the next 6-10 years.

Recommendations:

1. That the Australian government, under your stewardship, adopts the WSARA-2021 Affordability & Producibility Way as the new Defence industry best practice weapons Acquisition methodology for all future Defence Weapons Acquisition programs.
2. That the Australian Government also legislates how to EXECUTE the WSARA-2021 Legislation as described in Chapters 16, 17, and 18.
3. That sustainment requirements be mandated early in the Development cycle via a robust SRL type measure.

4. That the legislation mandates a new Gateway Project Management method as described previously.
5. That all IP rights issues are mandated during the initial contestability stage.

The end purpose of these recommendations and proposed legislation is to enforce a new type of compliance on both Defence and the Defence supply chain in ways that Defence and the supply chain cannot possibly avoid - as they have so very successfully avoided after every prior reform attempt in the past.

Deliverable Results from the Legislation:

- 20-40% **lower** total cost.
- 20-30% **less** total time.
- ~30% **lower** sustainment costs.
- More **affordable** equipment.
- More easily producible.
- More easily manufacturable.
- More exports.
- More innovation.
- Better technology.
- Better Warfighter capability.

What is Needed to Drain the Swamp:

1. We **must** create a new style of organisational bureaucracy based on a new World's Best Practice, highly competent, highly functional, highly effective, wholly fit for purpose, and wholly fit for office.
2. The Australian people must demand a much higher standard of competency and performance from all of those who serve us if we are to take our nation forward into the economic and social future we now face.
3. This solution to the Defence Swamp must deliver this new higher competency and performance requirement from each and every future Weapons System Acquisition program.
4. Nothing short of total reform of Weapons Acquisition is acceptable.

5. Nothing short of complete disruption of the Defence Swamp is acceptable.

In Conclusion:

Only deliberate concrete action by the Parliament can enforce a change upon Defence Weapons Acquisitions.

Only deliberate legislation specifically tasked to REFORM Weapons Acquisitions will work.

Only the Weapon Systems Acquisition Reform Act 2021 can drain the Defence Acquisition Swamp.

22

CONCLUDING HOW TO DRAIN THE SWAMP

This book only became possible after I decided to stop working with most new clients and wind down existing clients over the next few years. As an active consultant it was not possible to write this book. That's because I couldn't say the things I have said in this book about potential clients and then expect to get work from them.

But, freed from that constraint, I can now say what I really think about current practices within Defence and Industry.

My job in this book has been to lay out the issues and facts as I see them. Then paint the alternate realities as Defence sees them. And then let you the reader to decide for yourself where the best truth lies.

For me, this book represents some unfinished business.

There is no reason beyond our own ineptitude and incompetence why Defence Weapons Acquisition cannot become truly World Class.

There is no reason why our inept Weapons Development System cannot become World Leading.

There is no reason other than our own ineptitude why Australia cannot adopt World's Best Practice in the design and development of more AFFORDABLE and easily PRODUCIBLE Defence Weapons Systems.

There is no reason beyond our own ineptitude and incompetence why we can't drain the Defence Swamp.

So, the next time you hear Defence declare success you should very bluntly ask: "Success? Compared to what?"

Back in the Part 1 Introduction I wrote:

SWAMP: Current Usage - Government Bureaucracy of dysfunction, waste and incompetence, wholly unfit for purpose, and wholly unfit for office.

I think any reasonable reader would conclude that the current Defence Weapons Acquisition methodology and the 'alternate realities' as described in Part 1 is a Government Bureaucracy of dysfunction, waste and organisational incompetence, wholly unfit for purpose.

For far too long now Defence has been selling us 'square wheels' - and telling us they will eventually make them round. That time is over. The time for disruption has arrived.

In the Part 2 introduction I wrote:

SWAMP: A Solution - Government Bureaucracy of World's Best Practice, highly competent, highly functional, highly effective, wholly fit for purpose, and wholly fit for office.

The Solution to the Defence Weapons Acquisition Swamp is to execute a new WBP2020 type of thinking based on the Hybrid Stewart/Munro/Toyota Product Development System described previously as a full replacement for the existing dysfunctional methods which have been applied with such a 'foolish consistency' by Defence.

Success is about creating and establishing Superior SYSTEMS for Weapons Acquisition programs. Those Superior SYSTEMS described in Chapters 15 to 19 then deliver much needed better outcomes.

Those Superior SYSTEMS enable the Swamp to be drained.

Now is the time for a new Blueprint of Reform

A Blueprint for a reformed Defence Weapons Systems Acquisition Reform methodology where:

A new and brighter future where Affordability and Producibility is mandated.

A new and brighter future where the Taxpayer's hard-earned money is respected.

A new and brighter future where the Warfighter as the centre of the universe is respected.

A new and bright future freed altogether from the past.

Only a full root and branch change to the way Defence thinks about Weapons Acquisition can deliver the results that the Taxpayer and Warfighter demand from those who will deliver Weapons System capability in future.

Now is the time for WBP2020

WBP2020 is not just about a better Weapons Development System.

WBP2020 is about what Defence can do for the Taxpayer.

WBP2020 is about what the Weapons Systems supply chain can do for the Warfighter.

WBP2020 is about what your people can do for your own company using this same thinking.

The whole purpose of WBP2020 is the deconstruction of complexity into much simpler deliverable outcomes.

The whole purpose of the solutions described is to eliminate 'wicked problems' from Weapons Acquisitions.

We simply need to tap into the world's best practice thinking that the most astute of CEO's have already started.

Now is the Time for Reform

Chapter 19 described the sort of results and benefits from a REFORM of Weapons Acquisition programs.

These better results fully justify the use of the disruption model described in the preceding chapters.

These better results demand that the old ways must die a quick death.

These better results demonstrate a new method that is wholly fit for purpose.

These better results will deliver a 'draining of the defence weapons acquisition swamp'.

That will require a whole new mindset about how Defence thinks about Weapons Acquisition.

That will require a whole new leadership model for Weapons Acquisition.

That will require a clean-out of all of the 'dead-wood' that exists today.

That will require someone inside Defence with the cojones to REFORM Weapons Acquisition.

That will require someone to drive Weapons Acquisition REFORM in Defence backed up by Expert SYSTEMS and an Expert group of people orchestrating everything.

Finally, that persistently nagging question remains: Who has the cojones to totally REFORM Weapons Acquisition?

No-one should underestimate how difficult it will be to change the status quo in Defence Weapons Acquisition and the deep determination to retain the 'sacred cows' that exist today. The resistance will be fierce, persistent and sustained despite the apparent ease of implementing the solution as described in Chapters 15 to 19.

This warning should not be ignored by anyone.

EPILOGUE

**For some, this marks the END of a journey...
But for others, this is not the end, but a mere pause
... until Book Two arrives**

...where the Organisational Incompetence Theme Continues

Book Two broadens the same basic swamp theme 'a system wholly unfit for purpose' and applies it to Australia's completely dysfunctional Innovation System.

It's completely dysfunctional, and 'wholly unfit for purpose', simply because, despite spending **inputs** of around $30b on Innovation (about half of which is taxpayer money) our Innovation system **fails** to deliver any significant economic growth and jobs **outputs** from that $30b investment.

But it does not have to be this way.

Book Two will also lay out a Blueprint for reform of our Innovation system into a new far brighter future in which a now reformed Innovation system that is 'wholly fit for purpose', delivers strong economic growth and the strong creation of new and better Jobs.

... or until Book Three arrives

...where Organisational Incompetence is Resolved:

Book #3 deals directly with why nothing ever changes, not just in Weapons Acquisition (this book), or in book #2 which deals with why our national Innovation Systems fail, but why the systemic organisational failures from incompetent leadership, and the resultant organisational incompetence, are so deeply pervasive across all aspects of our Political, Bureaucratic, and Industry leadership landscapes.

But it does not have to be this way.

Book Three will also lay out a Blueprint for complete reform of our Political Bureaucracy system into a new far brighter future in which a now reformed Bureaucracy system of organisational competence that is 'wholly fit for purpose', finally delivers on the promise of strong economic growth and jobs, while creating better societal outcomes for Australia.

APPENDIX A - OTHER LINKAGES

While this book is very specifically focussed on Defence Weapons Acquisition problems, in reality the cost and time blowouts, wicked problems, highly dysfunctional outcomes and just plain dumb decisions described in this book also apply across most of Industry, the corporate world and other government functions or Qangos.

I am sure that astute readers will realise that the content in this book has far wider reform application across wide expanses of the economy. The content from both halves of this book can apply just as equally to all types of industries and firms involved in Innovation, Product Design, Manufacturing or the delivery of Services.

I am sure many of you experienced your own sense of déjà vu as the book unfolded because it is not unique to Defence, but can apply everywhere such is the universal nature of the problems described in this book

I am sure many of you can think of places where the Hybrid Stewart/Munro/Toyota Product Development System could be applied to drive massive reform when compared to existing failed or failing methods.

As a simple example I will describe four industries that need similar reform:

Rail:

Rail suffers from most of the same problems as Defence.

In both Queensland and NSW, they have built trains that don't fit platforms. In Queensland, they built new country trains with toilets that were too small. In NSW, they built new country trains that don't fit through the tunnels under the Blue Mountains. In Victoria, they built trains that are too heavy and wear out the wheels. In Victoria, they built new tracks with radius bends that are too sharp meaning the country trains have to slow to a running pace around these bends.

All of these failures have been, or need to be, remediated at taxpayers' expense.

Major Infrastructure:

Who among us has not heard of major and ongoing problems in Road or Rail infrastructure programs?

Huge cost blowouts, long time delays, commuter frustration...

Road Tunnels in Sydney and Brisbane: collapsed due to faulty projections.

Melbourne East-west link: political needs cancelled the road - $1b compensation paid to deliver no road.

Sydney's new trams: $1.2b over budget and 2 years late.

NBN: A program originally slated to cost less than $20b, will end up costing more than $50b

All ultimately paid for by the Taxpayer.

Building and Construction:

Opal Tower (Sydney): New building from which everyone was evacuated because of potential building collapse.

Repairs took one year (remediation costs unknown, and the project is tied up in legal proceedings).

Flammable cladding: Tens of thousands of buildings across Australia are impacted.

The Taxpayer will end up paying for most of these blunders.

Hospitals:

Royal Adelaide Hospital: Said to be the most expensive building in Australia (even more expensive that our lavish casinos). $2.3b - 17 months late, and $600m over budget (but which budget?)

Parts of the building are considered not fit for purpose and may need to be rebuilt.

Perth Children's Hospital: Original budget $850m, final cost $1.2 b.

4 years late according to the original plan, 3 years late from the 'adjusted' plan.

The hits to the Taxpayer just keep coming.

I am sure every reader can think of many more examples of similar problems from their own personal experiences across other areas of Government or industry.

But more positively, each of the examples above, in turn describes the need for a similar type of reform program as that described in Chapters 16 to 19 of this book.

The need is great. Is the will there to meet the need and take action?

APPENDIX B - AUGUSTINE'S LAWS (1984)

Extracted from Wikipedia.[1]

Augustine's Laws are a series of tongue in cheek aphorisms (a pithy observation which contains a general truth) put forth by Norman Ralph Augustine, a luminary of the US aerospace industry, who served as US Under Secretary of the Army from 1975 to 1977.

In a book published in 1984, Augustine drafted his Droll views in a series of light-hearted 'laws' as humorous or sarcastic statements expressed in a mock serious manner that may have the reader mentally shivering at the very real possibility that these laws are in fact absolutely true.

For example: In one aphorism, he plotted the exponential growth of unit cost for fighter aircraft since 1910 and extrapolated it to its absurd conclusion ... Nearly three decades later, (1990's) he said, "we are right on target."

His most cited law is number 16, which shows that defence budgets grow linearly but the unit cost of a new military aircraft grows exponentially:

"In the year 2054, the entire defense budget will purchase just one tactical aircraft. This aircraft will have to be shared by the Air Force and Navy 3½ days each per week except for leap year, when it will be made available to the Marines for the extra day."

A Few Other Notable Laws:

Law Number VIII: The most unsuccessful *four years* in the education of a cost-estimator is fifth grade arithmetic.

Law Number XII: It costs a lot to build bad products (*my personal favourite.*)

Law Number XIV: After the year 2015, there will be no airplane crashes. There will be no take-offs either because electronics will occupy 100 percent of every airplane's weight.

Law Number XVII: Software is like entropy. It is difficult to grasp, weighs nothing, and obeys the Second Law of Thermodynamics: i.e. it always increases.

Law Number XVIII: It is very expensive to achieve high unreliability. It is not uncommon to increase the cost of an item by a factor of ten for each factor of ten degradation accomplished.

Law Number XIX: Although most products will soon be too costly to purchase, there will be a thriving market in the sale of books on how to fix them.

Law Number XXIV: The only thing more costly than stretching the schedule of an established project is accelerating it, which is itself the most costly action known to man.

Law Number XXVI: If a sufficient number of management layers are superimposed on each other, it can be assured that disaster is not left to chance.

Law Number XXXII: Hiring consultants to conduct studies can be an excellent means of turning problems into gold: your problems, into their gold.

Law Number XXXIV: The process of competitively selecting contractors to perform work is based on a system of rewards and penalties, all distributed randomly.

Law Number XXXV: The weaker the data available upon which to base one's conclusion, the greater the precision which should be quoted in order to give the data authenticity.

Law Number XXXVII: Ninety percent of the time things will turn out worse than you expect. The other 10 percent of the time you had no right to expect so much.

Law Number XXXIX: Never promise to complete a project within six months of the end of a year, in either direction.

Law Number XL: Most projects start out slowly, and then sort of taper off.

Law Number XLII: Simple systems are not feasible because they require infinite testing.

Law Number XLIV: Aircraft flight in the 21st century will always be in a westerly direction, preferably supersonic, crossing time zones to provide the additional hours needed to fix the broken electronics.

Law Number XLVIII: The more time you spend talking about what you have been doing, the less time you have to spend doing what you have been talking about. Eventually, you spend more and more time talking about less and less until finally you spend all your time talking about nothing.

These witticisms may also seem to carry all that much more weight because at the time Augustine was President of Martin Marietta Corp - a precursor to today's Lockheed Martin, where he also served as President (1995-1997).

---

1. *https://en.wikipedia.org/wiki/Augustine%27s_laws*

APPENDIX C - SUPERIORITY: ARTHUR C CLARKE (1951)

Superiority is a science fiction short story of 4000 words by British writer Arthur C. Clarke, first published in 1951.[1]

This story could have been written today as a metaphor for the F-35 Joint Strike Fighter Program, as well as other 'one-size-fits-all' or 'bewitching' high-cost weapon systems. (Augustine's Laws say much the same).

Clarke's story depicts an arms race during an interstellar war.

It shows the dominant side, which is far more technologically advanced, being defeated despite its apparent superiority, because of its easy willingness to discard old tried and true technology before fully perfecting the new 'bewitching' weapons.

Meanwhile, the inferior enemy steadily built up a far larger arsenal of tried and true weapons which, while far more primitive, were also far more reliable.

This fantastic and haunting short story warns us about the opportunity cost of hyper-focusing on developing super high-end weaponry, instead of improving and perfecting the existing technology first.

To read Clarke's original article in full visit - http://www.mayofamily. com/RLM/txt_Clarke_Superiority.html

In - When Superiority Goes Wrong Science Fiction and Offset Strategies – by Ben FitzGerald writes:

Clarke's short story, Superiority, does not predict technologies that we recognize today, but elegantly describes a number of disturbingly familiar military technical failure modes. Such insights are especially helpful when thinking about new endeavours like the Pentagon's Defense Innovation Initiative, which will include both a new long-range research and development planning program and an offset strategy.

In a distant future, an unnamed dominant military power has been engaged in a lengthy space war with a technically inferior adversary. The dominant force appoints a new 'Professor General' theoretical scientist to fill a post of vital importance.

This new leader changes the dominant power's technology strategy from upgrading existing systems incrementally to developing and deploying entirely new weapons, believing that "a revolution in warfare may soon be upon us." This change in strategy sets off a series of disastrous events that ultimately leads to the dominant military power's defeat.

Here's how the decline unfolds. The superior force abandons the production of old weapons platforms to focus on the development of a new 'irresistible weapon.' The weapon takes longer to develop than planned and can only be launched in limited quantities. During the development period, the adversary is able to build larger numbers of their inferior weapons so that even when the new weapon works as planned, it does not provide the anticipated advantage. The superior force then attempts a large-scale effort at battle management automation only to have the enemy rapidly adapt to their new concept of operations, targeting central nodes in their new order of battle to devastating effect. In response, the previously superior force develops a final new weapon only to have significant integration issues that throw their own forces into total disarray, precipitating their defeat within a month.

The story hits home for all of us familiar with the challenges of developing new military technologies and capabilities. Many of these issues can be seen to lesser degrees in recent air, sea, and land weapons system developments. Clarke's story serves as a powerful reminder that these issues should not be attributed to technology, the cunning of our adversaries, or macro trends in technology diffusion. These are ultimately human failings.

The United States relies on technical superiority to maintain its military advantage. But this technical superiority requires humans to generate the right strategies, design and build the right technologies, devise concepts of operations, and train forces to operate the technology to achieve strategic and tactical objectives. Sometimes this requires new 'leap ahead' technology and sometimes it does not. Given that human judgment is required, and that all humans are fallible, we cannot hope to be right 100 percent of the time.

However, we cannot ever let the hubris evident in Superiority lead us to defeat due to, as the narrator assesses, "...the inferior science of our enemies."[2]

---

1. *http://www.mayofamily.com/RLM/txt_Clarke_Superiority.html*
2. *https://warontherocks.com/author/ben-fitzgerald/*

CONTACT DETAILS

Volumes 2-4 of the Drain the Swamp Series will be released early in 2021.

Visit www.DraintheSwampbooks.com and sign up to be notified as each new volume is released so you can be the first to read this brutally honest picture of the current state of Weapons (and other government) Acquisitions and discover the solutions that are so urgently needed - and already available.

Contact Gary via the website.

ABOUT THE AUTHOR

Gary D. Stewart, Author of the 'Drain the Swamp' Books

Gary D. Stewart has decades of experience in transforming complex business systems encountering first-hand the unaffordability of incompetent product design and development processes then mastering and teaching reproducible excellence and competency in this area.

He worked at the world's most effective company alongside the leading thinkers in this field.

Gary's style is direct, his honesty compelling. "If your goal is to make the complex simple and the simple compelling, you need to bluntly identify problems, ruthlessly eliminate them, and - above all - rigorously implement best practices in a competent manner."

www.ingramcontent.com/pod-product-compliance
Lightning Source LLC
Chambersburg PA
CBHW021925190326
41519CB00009B/918